The AI Efficiency & Process Optimization™

How Leadership Teams Convert AI Adoption Into Measurable Operating Performance and a Defensible Financial Return

Stephen R. Jordan

SRJ Consulting & Services Publishing

Copyright

The AI Efficiency & Process Optimization™

How Leadership Teams Convert AI Adoption Into Measurable Operating Performance and a Defensible Financial Return

The Operating Discipline for AI Library™, Volume IV

Pillar I: AI Business Services™

Volumes are numbered by operating sequence, the natural order in which the disciplines build inside a business.

Copyright

Published by SRJ Consulting & Services LLC, Frisco, Texas. SRJConsultingServices.com.

Names: Jordan, Stephen R. (Stephen Ray), 1960- author.

Title: The AI efficiency & process optimization : how leadership teams convert AI adoption into measurable operating performance and a defensible financial return / Stephen R. Jordan.

Other titles: Artificial intelligence efficiency and process optimization

Description: First edition. | [Frisco, Texas] : SRJ Consulting & Services Publishing, [2026] | Series: Operating discipline for AI library ; v. 4. | Includes bibliographical references.

Identifiers: ISBN: 9798996940219 (Paperback) | 9798996565894 (Hardback) | 9798996940202 (Digital)

Subjects: LCSH: Artificial intelligence--Management. | Business enterprises--Data processing--Management. | Business enterprises--Decision making. | Information technology--Management. | Industrial productivity. | Productivity accounting. | Senior leadership teams. | Leadership. | BISAC: BUSINESS & ECONOMICS / Information Management. | COMPUTERS / Artificial Intelligence / General. | BUSINESS & ECONOMICS / Management.

Classification: LCC: HD30.2 .J674 2026 | DDC: 658/.0563--dc23

Trademarks

mance Governance™, AI Performance Scorecard™, AI Readiness Attestation Sheet™, AI Readiness Maturity Scale™, AI Refinement Register™, AI Risk & Governance Review™, AI Risk Governance & Security™, AI ROI Evaluation Framework™, AI Stakeholder Communication Matrix™, AI Steering Committee Charter™, AI Third-Party Governance Statement™, AI-to-Outcome Alignment Map™, AI Use Case Custodian Log™, AI Vendor Risk Inventory™, AI Adoption Pattern Map™, Application Security in the Age of AI™, Approved AI Tool Register™, Baseline Verification Standard™, Bi-Annual AI Readiness Reassessment Log™, Cloud and Infrastructure Security in the Age of AI™, Data Accountability and Remediation Checklist™, Data Reliability Checklist™, Decision Control Matrix™, Decision Influence Matrix™, Expand-Refine-Pause Decision Protocol™, Expansion Authorization Profile™, Foundation Pause Mandate™, Governance Communication Cascade™, Master 90-Day Execution Matrix™, Master AI Readiness Scorecard™, NEYR™ (Net Efficiency Yield Ratio), OLF™ (Operational Leakage Factor), Operational Health Check™, Operational Integration & Workflow Adoption™, Performance Reality Test™, Refinement Remediation Protocol™, Repeatable Governance Framework™, Secure by Design in the Age of AI™, Standing AI Adoption Policy™, The 6-Step Review Process™, The AI Business Enablement Audit™, The AI Efficiency & Process Optimization™, The AI Efficiency Gap™, The AI Efficiency Scorecard™, The AI Efficiency Tax™, The AI IT Security Audit™, The AI IT Security Implementation & Strategy™, The AI Operating System™, The AI Readiness & Performance Assessment™, The AI ROI Formula™, The Four Audit Outputs™, The Operating Discipline for AI Library™, Weekly AI Operational Spot Check Log™, Workflow Readiness Review™, 90-Day AI Decision Action Plan™, 90-Day AI Readiness Roadmap™, and 90-Day Governance Launch Plan™. All other trademarks referenced are the property of their respective owners.

Disclaimer

or independently verified financial information without separate validation by qualified financial professionals.

Case examples in this book are composites synthesized from patterns observed across many consulting engagements spanning more than two decades of professional practice. The specific numeric details (firm size, industry, workflow specifics, financial outcomes, time horizons) are illustrative constructions designed to convey operating patterns in concrete terms. No single composite represents a single real engagement; every composite combines elements from multiple distinct situations. Any resemblance to a specific organization, engagement, or client is unintended and coincidental.

Neither the author nor the publisher shall be liable for any loss of profit or any other commercial damages, including but not limited to special, incidental, consequential, or other damages, arising directly or indirectly from the use or application of any information contained in this work.

AI Disclosure

This book was produced with AI assistance. The text was developed through extensive use of AI tools, with substantial editorial review, structural revision, and final approval by the author. All figures, diagrams, scorecards, and illustrations were generated using AI tools under the direction of the author, who reviewed and approved each one for accuracy and appropriateness. The author retains full responsibility for the content, analysis, frameworks, judgments, and conclusions presented in this work. The book is an original English-language work; no portion was machine-translated from another language.

Cover design by Muhammad Shakir Tahir, Book Bloom Studio.

First Edition

Dedication

For the executives who built the AI program, defended its budget, governed its risks, and now have to prove it produced a result. The numbers in this book are how that proof gets built.

Acknowledgments

This book is the fourth Volume of a working library that began with the AI Business Enablement Audit™, continued with the AI Readiness & Performance Assessment™, and built the governance record through the AI Risk & Governance Review™. Each Volume has drawn on the same foundation: more than two decades of operational practice across accounting, enterprise technology, cybersecurity, and consulting engagements with leadership teams running real businesses under real pressure.

The efficiency and process optimization material in this Volume reflects more than two decades of operational practice and consulting work with leadership teams confronting the gap between AI activity and AI evidence. The patterns described in this book emerged through extended observation of how AI operates inside mid-market businesses, and the framework was developed by abstracting recurring patterns into operating instruments that any leadership team can apply. Their willingness to look at what AI was actually producing, rather than what it appeared to be producing, is what made the framework possible. The case patterns appearing in these pages are composites, drawn from those engagements with identifying details abstracted.

The measurement methodology, the loaded labor rate work, the AI Efficiency Tax framing, and the financial integration discipline in this Volume are the product of careful work with CFOs, controllers, and finance leaders who pressure-tested the calculations against their own models. Particular thanks to the colleagues who refused to accept activity metrics as performance metrics, and whose insistence on tying every AI claim to a defensible number shaped the standards in this Volume.

The case continuity across the four Volumes is deliberate. Readers who recognize the 75-person construction firm, the 40-person accounting firm, and the 60-person professional services practice from earlier Volumes will see them again here. The composite patterns are continuations of the same operating realities those businesses face as AI work moves from visibility (Volume I) to readiness (Volume II) to governance (Volume III) to optimization (this Volume).

Thanks are due to the executive peers and operating colleagues who challenged the framework before it reached publication, the consulting clients who tested its earliest drafts in working engagements, and Shakir, whose visual judgment has shaped how this Volume reaches its reader.

The errors that remain are mine.

Stephen R. Jordan

Frisco, Texas, 2026

Executive Summary

Volume I gave you visibility into the AI running inside your business. Volume II gave you a defensible decision about whether to expand, refine, or pause each use case. Volume III built the governance record that defends those decisions to any stakeholder who asks. None of those three Volumes answered a different question that has now arrived in nearly every leadership room I work in: now that AI is operating inside the business, what has it actually produced in numbers you can defend?

That is the question this book is built to answer.

The AI Efficiency & Process Optimization™ is the fourth operating instrument in the library and the closing Volume of Pillar I, AI Business Services™. It sits inside the AI Operating System™ that Volumes I, II, and III installed. It does not replace any framework already in place. It extends the AI Tool Inventory from Volume I. It extends the readiness scores and the workflow readiness records from Volume II. It extends the per-use-case governance dossiers and the AI Corrective Action Register from Volume III. And it adds the new instruments that define this Volume: the AI Efficiency Gap, Phantom Productivity, the AI Efficiency Tax, the Workflow Reality Map, the AI Process Fit Test, the AI Efficiency Scorecard, the AI ROI Formula, the Executive AI Efficiency Brief, and the 90 Day AI Process Optimization Plan.

Three forces have made operating performance personal for executives in the last eighteen months, and none of them are going away. The CFO has arrived at the AI conversation. Budget reviews now include line-item AI spend, and the AI line increasingly has to defend itself against the same return-on-investment standards the business applies to every other capital allocation. The board has arrived at the AI conversation. Quarterly operating reviews now include questions about what AI has produced for the business, and a sense that adoption is happening is no longer sufficient as an answer. The lender, the investor, and the acquirer have arrived at the AI conversation. Diligence now includes AI operating performance, and a business that can document measurable improvement supports a higher valuation while a business that cannot document anything supports a discount.

This book is not a measurement manual. It is an operating book. It does not require a Big Four firm, a Chief AI Officer, or a dedicated analytics department. It requires the same operating discipline a leadership team already applies to finance, hiring, and customer obligations, and it gives that discipline the structure to produce the financial answer the business now needs to have ready.

Contents

Introduction

You Made the Bet. Now It's Time to Prove It Paid Off

You spent the money. You told your team AI was going to change the business. You pushed through the friction of adoption, dealt with the skeptics in the room, and built enough structure to know that AI is no longer just an experiment inside your company. It's running inside real workflows, touching real outputs, and consuming a real line on your budget every single month.

Now someone is asking the question.

It might be your CFO looking at the AI spend during a budget review. It might be a board member who wants more than enthusiasm before the next approval. It might be a lender, an investor, or an acquirer who wants to see what AI has actually done to your operating performance before they make a decision. Or it might just be you, sitting alone with a spreadsheet at the end of a quarter, trying to figure out whether any of this has moved the needle in a way you can actually defend.

The question is always some version of the same thing: what has AI actually done for this business?

Not what tools are being used. Not how many employees have logged in. Not how often the team mentions AI in meetings or how polished the outputs look. The real question is whether anything measurable has changed. Whether a workflow is faster. Whether a cost is lower. Whether an error rate has dropped. Whether capacity has opened up. Whether margin has held or improved because of something AI did that the business can trace back to a specific process and a specific number.

Most leaders don't have a clean answer to that question right now. And that gap between what AI appears to be doing and what it's actually doing to the P&L is exactly what this book is built to close.

Volumes I, II, and III of The Operating Discipline for AI Library brought you to this point. Volume I created visibility into where AI was operating inside the business. Volume II tested readiness and helped leadership make structured decisions about which AI use cases were worth expanding. Volume III built the governance record that defends your AI decisions to any stakeholder who asks. That work was essential. It gave the business a foundation that most organizations don't have. But it was always leading somewhere. It was always building toward the executive question that governance alone can't answer.

This volume answers that question.

Volume IV is not about AI adoption. It's not about getting more tools into more hands or building more enthusiasm inside the organization. It's about taking the AI use cases already running inside the business and determining, with real numbers, whether they're producing measurable operational improvement or quietly generating cost, rework, and management overhead that nobody is tracking. It's about converting the audit findings, readiness decisions, and governance records from the first three volumes into a structured optimization process that produces results you can see in cycle time, capacity, error rates, labor savings, margin, and ROI.

That's the work of this book. And it starts now.

What You Are Going to Learn and Why This Book Is Different

There's no shortage of books about AI productivity. Most of them follow the same structure: here are the tools, here are the prompts, here's what AI might someday do for your business. They're written for people who are still deciding whether to adopt AI. That's not you. You've already adopted it. You're past the enthusiasm phase. You're in the accountability phase, and that phase requires a different kind of book.

This is an operational execution book for the leader who is accountable for AI results, not just AI activity. The difference between those two things is bigger than it sounds. AI activity is easy to generate. You can fill a dashboard with usage statistics, output volumes, and adoption rates in a matter of weeks. (Solomon et al., 2024) AI results are harder. They require workflow mapping, baseline measurement, structured testing, and a clear line connecting what AI did to what the business saved, gained, or improved. (Wamba, 2020) Most organizations have the first and are missing the second entirely.

You will learn how to map the workflows AI is already touching, identify where it is creating real efficiency and where it is creating the appearance of efficiency, measure cycle time, capacity, error rates, and labor savings with numbers you can stand behind, and calculate an AI return on investment that holds up under CFO scrutiny. You will also learn how to take the outputs from the first three Volumes (the AI Tool Inventory, the readiness decisions, the governance records) and feed them directly into a 90 Day AI Process Optimization Plan that fits inside your existing operating rhythm. No new infrastructure, no data science team, no enterprise-scale overhaul. A structured, sequenced plan that a lean leadership team can execute with the resources it already has.

What makes this book different is simple. Every chapter connects AI to the P&L. Every tool is built for a small or mid-sized business with a real budget and a lean team. And every framework is designed to produce something you can walk into a room with and defend, not just something you can read and feel good about. That is the standard this book is held to. That is the standard you should hold it to as well.

The Road Map. What This Book Covers and the One Lesson That Changes Everything

This book moves from diagnosis to execution. It does not start from zero, and that distinction matters. Most operational improvement books assume you are building from scratch. This one assumes you have already done the foundational work. You have visibility into your AI use cases. You have made readiness decisions. You have a governance record. The question now is what to do with all of that to produce measurable improvement.

The road map begins with the AI Efficiency Gap, the distance between what AI appears to be doing inside your business and what it is actually doing to your operating performance. Understanding that gap, measuring it, and closing it is the central task of this book. Everything else is built around it.

From there, the book moves through Phantom Productivity, what happens when AI creates the appearance of efficiency without the substance of it. Outputs look polished, turnaround times look faster, and the team looks productive, but rework is accumulating, review burden is growing, and the actual cost of getting work done has not dropped. In some cases it has gone up. (Lipscomb, 2011; Brynjolfsson et al., 2018) Most leaders do not realize it is happening until they start measuring the right things.

Then comes the AI Efficiency Tax, the hidden operational cost of reviewing, correcting, and managing AI outputs. This cost is almost never tracked as a separate category. It gets absorbed into labor hours, management time, and rework cycles nobody connects back to AI. The Tax is quiet, cumulative, and directly corrosive to margin. (Zhang, 2025) This book gives you a way to find it, quantify it, and eliminate it.

The Workflow Reality Map shows what an AI-supported process actually looks like rather than what it was designed to look like, where the real handoffs happen, where errors enter, and where review burden has accumulated. (Friebe, 2026) The AI Process Fit Test gives you a structured way to evaluate any active AI use case and produce an explicit decision: expand, refine, pause, or remove.

The Four AI Performance Indicators (cycle time, capacity, error rate, and rework cost) are the four numbers that prove whether AI has actually moved the operation. The AI Efficiency Scorecard converts those four indicators into a structured operating record across every active use case. The AI ROI Formula takes the Scorecard data and converts it into a defensible return on investment number a CFO, a board, a lender, or an acquirer can examine without it falling apart. The Executive AI Efficiency Brief gives you a one-page communication format that replaces the AI enthusiasm update with an AI performance report. The 90 Day AI Process Optimization Plan pulls everything together into a sequenced execution plan that runs inside your existing operating rhythm without requiring a separate project team.

The lesson that runs through every one of these tools is the same, and it is worth stating clearly before you read another page. AI activity is not AI efficiency. Usage is not improvement. Adoption is not optimization. (Umoga et al., 2024) A business that cannot measure what AI has moved is still managing a story rather than a result. A story, no matter how well told, will not hold up in a board meeting, a lender review, or an acquisition conversation when someone asks for the numbers behind it. By the time you finish this book, you will have the language, the measurements, and the operating sequence to replace the story with evidence.

Who This Book Is For and How to Use It

This book is written for owners, CEOs, CFOs, COOs, managing partners, board members, lenders, investors, consultants, and operating leaders in small and mid-sized businesses that are already using AI and now need to prove what it has actually improved. If you are responsible for the P&L, accountable for operating performance, or in a position where someone is going to ask you to justify your AI investment, this book is for you.

It assumes you are not looking for another explanation of what AI can do. You already know. You have seen the demos, read the articles, and heard the pitches. What you need now is a practical, structured way to measure what AI has done inside your specific workflows, your specific labor model, and your specific margins. That is a different problem, and it requires a different kind of book.

To get the most value from this Volume, read it with your current AI Tool Inventory, your readiness decisions from Volume II, your governance records from Volume III, and your top three AI-supported workflows in front of you. Each chapter is designed to help you test one part of your AI operation, document what you find, and convert that finding into a decision your leadership team can act on. The tools in this book are not meant to be read and set aside. They are meant to be used, filled in, and brought into your next leadership meeting.

If you have not completed Volumes I, II, and III, this book still works. You will need to do some foundational mapping as you move through the early chapters, and the book will guide you through that. If you have completed the earlier Volumes, you are starting from a significant advantage. You already have the AI Tool Inventory that feeds the Workflow Reality Map. You already have the readiness decisions that inform the AI Process Fit Test. You already have the governance records that support the AI ROI Formula. This Volume takes that existing work and converts it into measurable operational improvement that can be seen in the business, defended in the boardroom, and reflected in the P&L.

Where Volume IV Sits in the Library

Volume IV is the closing Volume of Pillar I, AI Business Services™. The four Volumes in Pillar I sequence the operating disciplines a business needs to install AI honestly: visibility (Volume I), readiness (Volume II), governance (Volume III), and optimization (Volume IV). Volumes I, II, and III build the foundation that makes AI defensible. Volume IV builds the layer that makes AI valuable. A business that finishes Pillar I has both.

Pillar II, AI Risk Governance & Security™, runs in parallel and addresses the security side of the AI Operating System™ through five further Volumes. You do not need to wait for Pillar II to finish this Volume's work. The two pillars are deliberately independent. A full catalog of all nine Volumes appears in the back matter under Continuing the Work.

Read this Volume with the artifacts from the prior three on the table next to you. The AI Tool Inventory from Volume I. The readiness scores and workflow readiness records from Volume II. The per-use-case governance dossiers and the AI Governance Maturity Scale scores from Volume III. Those documents are not background context. They are the named inputs this Volume converts into operating improvement. The closer they are to your hand while reading, the faster you will see where each Volume IV tool fits and what it requires from you to produce the result it promises.

The place to start is understanding exactly why the gap between AI activity and AI efficiency exists, how wide it already is inside your business, and what it costs you every single month it stays open.

Chapter 1: The AI Efficiency Gap™

Why Adoption Is Not the Same as Improvement

CHAPTER 1 OUTCOMES

At the end of this chapter, you will be able to:

1. Name exactly what the AI Efficiency Gap is, why it appears, and how to spot it inside your business.
2. Identify which of three conditions (missing baseline, unmapped workflow, activity-only measurement) apply to your AI use cases.
3. Recognize why standard KPIs fail to surface the gap and why workflow-level measures are required.
4. Explain how Volumes I through III feed Volume IV — turning audit, readiness, and governance outputs into optimization inputs.
5. Commit to a first measurement: name three high-activity AI workflows and walk through the four diagnostic questions.

ACTIVITY IS NOT EVIDENCE. WORKFLOW-LEVEL MEASUREMENT IS WHERE THE GAP GETS CLOSED.

The Problem This Chapter Solves: You Have AI Activity Without AI Evidence

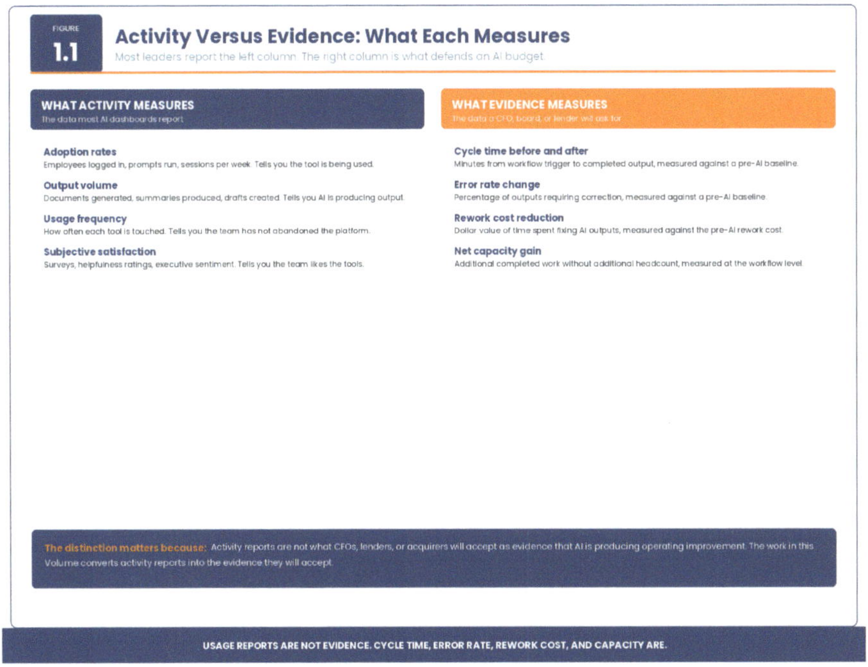

Most leaders can tell you who is using AI. They can name the tools, identify the departments with the highest adoption, and pull a usage report that shows prompts run, documents generated, and hours logged. That information feels like progress. It gets reported in leadership meetings as evidence that the AI investment is working. And in most organizations, nobody pushes back on it because the activity looks real and the numbers look good.

But activity and evidence aren't the same thing (Narayanan & Kapoor, 2025).

The harder question, the one that almost never gets asked in those same meetings, is whether anything measurable has actually changed. Not whether AI produced more output. Not whether employees are logging in more often. Whether a workflow is faster. Whether a cost is lower. Whether an error rate dropped. Whether a capacity constraint was relieved. Whether a margin was protected because of something AI did that the business can trace back to a specific process and a specific number.

If the honest answer to those questions is "we think so" or "our team says it's helping" or "we're pretty sure it's saving time somewhere," then the business has an AI Efficiency Gap, and it's almost certainly costing more than anyone on the leadership team realizes right now.

That gap is what this chapter is about. Not the technology. Not the tools. The gap between what AI appears to be doing and what it's actually doing to the business's operating performance. The gap between adoption and improvement. Between a tool being used and a result being produced.

The work that opens this Volume is naming that gap clearly, explaining why it appears even in businesses that have already done the adoption, readiness, and governance work, and giving you a way to locate it inside your own operation before moving into the optimization work that follows. Every chapter in this book depends on understanding this gap first, because you can't close something you haven't measured, and you can't measure something you haven't named (Boediman & Ferdiana, 2025) .

The AI Efficiency Gap is the first thing to name. And once you see it clearly, you won't be able to unsee it.

Defining the AI Efficiency Gap™. The Distance Between What AI Promised and What It Delivered

The AI Efficiency Gap is the measurable distance between what leadership expected AI to improve and what the business can actually prove AI has changed. It shows up in a specific way. Cycle times stay flat. Error rates don't move. Capacity pressure stays the same or gets worse. Rework quietly increases. The P&L shows no meaningful improvement even though AI usage has grown steadily for months. The tools are running. The team is busy. But the numbers that matter haven't moved.

Most leaders who encounter this pattern assume the tools aren't working. That's usually not the problem.

The gap almost always comes from three causes working together. First, the business never captured a baseline before AI was introduced, so there's no reference point for measuring what changed. Second, AI was placed inside workflows that were never mapped or standardized, so the tool is operating inside a process that nobody fully understands. Third, the organization measured activity instead of operational results, so the dashboards look full while the underlying performance stays stuck.

When all three of those conditions are present at the same time, the AI Efficiency Gap is inevitable. The tools can be excellent. The team can be enthusiastic. The governance can be solid. But without a baseline, without a mapped workflow, and without the right measurements, there's no way to know whether AI is actually improving anything. The business is managing presence, not production.

How the Three Conditions Combine to Create the Gap

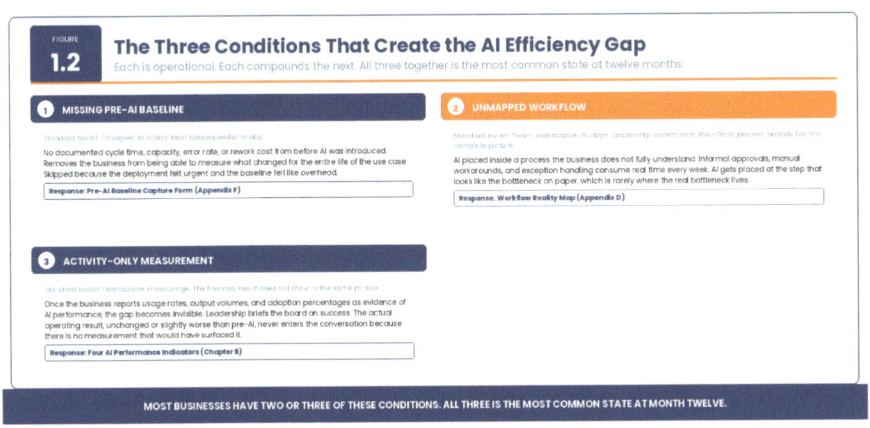

The three causes named above, missing baseline, unmapped workflow, and activity-only measurement, do not all carry the same operational weight. They are not equal contributors. They are sequential, and the order in which they appear inside a business determines how expensive the gap becomes before anyone notices it.

The missing baseline comes first. It is the cheapest condition to install and the most consequential one to skip. Capturing a pre-AI baseline takes a day of work for most workflows. Skipping it removes the business from being able to measure what changed for the entire life of the use case. Most leadership teams skip it because the AI deployment feels urgent and the baseline feels like overhead. The cost of that decision compounds for as long as the workflow runs without measurement, which in practice is until someone in a board meeting asks a question the leadership team cannot answer.

The unmapped workflow comes second. When AI is introduced into a process that the team has never formally walked through end to end, the tool is doing work inside a system the business does not fully understand. The team may understand its own steps. The leadership team may understand the official process. But nobody has the complete picture that includes the informal approvals, the manual workarounds, and the exception handling that consume real time every week. Without that picture, AI gets placed at the step that looks like the bottleneck on paper, which is rarely the step where the real bottleneck lives.

Activity-only measurement comes third and is what locks the gap in place. Once the business is reporting AI usage rates, output volumes, and adoption percentages as evidence of AI performance, the gap becomes invisible. The dashboard says the workflow is thriving. The team confirms it. Leadership briefs the board on the success. And the actual operating result, which is unchanged or slightly worse than the pre-AI state, never enters the conversation because there is no measurement that would have surfaced it.

Each condition compounds the next. A missing baseline makes the unmapped workflow harder to fix, because there is no reference point for what the workflow used to produce. The unmapped workflow makes activity-only measurement more dangerous, because the activity is happening inside a process the business does not understand. The three together produce the AI Efficiency Gap at maximum width, which is exactly the state most businesses are in twelve months into AI adoption, and exactly the state this Volume is built to correct.

The Cost of Leaving the Gap Open Another Quarter

There is a specific argument that leadership teams use to defer the work in this Volume that is worth addressing directly. The argument runs like this: AI is still developing. The tools are improving. The team is still learning. Measuring performance now is premature, because the numbers will be unreliable and the workflow will look worse than it actually is. Better to wait another quarter or two until the operation settles, then start measuring.

That argument is wrong, and waiting another quarter is the most expensive choice the business can make.

The waiting argument confuses tool maturity with workflow maturity. The tool is developing. The vendor will release updates. The model will improve. None of that changes how the workflow around the tool is performing today. The workflow has a current cycle time, a current rework cost, a current capacity profile, and a current AI Efficiency Tax. Those numbers exist right now whether or not the business is measuring them. Waiting does not improve them. It only delays the moment when the business learns what they are.

The cost of the delay is quantifiable. If a workflow is running an AI Efficiency Tax of three thousand dollars per month, every quarter the business waits to measure costs nine thousand dollars in margin leakage that could have been recovered. Multiply that by three or four use cases running unmeasured, and the deferral cost reaches the range where it would have funded a full-time hire by year end. The business is paying for the gap whether or not it has named the gap. Naming it is the first step toward stopping the payment.

The right response to the waiting argument is to measure now, document the measurement honestly, and let the numbers improve through optimization rather than through delay. A baseline captured during a difficult quarter is more useful than no baseline at all, because it gives the business a reference point for measuring whatever happens next. The numbers will move. The direction of movement is what the AI Operating System exists to track and improve. Standing still and waiting for the picture to look better is not a strategy. It is the absence of one.

Picture what this delay actually produces. A 40-person accounting firm decided to wait until tax season ended before measuring its AI return. The wait ran nine months instead of three. The CFO who finally asked for the numbers received a partial estimate and a list of caveats. Two senior partners had spent those nine months absorbing AI review burden into their billable hours, which meant client realization rates were quietly compressing. A junior staff member had built a personal prompt library that produced significantly better results than the firm's standard, and she left for a competitor. The AI return calculation, when it was finally built, was negative for the year. The recovery work took two more quarters. The measurement that would have taken a day before AI was introduced, and a week six months into the rollout, took a quarter once the gap had compounded across multiple use cases. Waiting was the most expensive decision the firm made.

To locate the gap inside your own company, start with a simple exercise. List the three workflows where AI is most active right now. For each one, ask whether you have a pre-AI baseline for cycle time, error rate, rework volume, or cost per output. Ask whether the workflow was formally mapped before AI was introduced or whether AI was dropped into an existing process without a structured review. Ask what measurement, if any, is currently tracking whether AI has improved that workflow's performance. If the answers are mostly "no" or "we don't know," the gap is open and it's sitting inside those workflows right now.

The key insight here is important to hold onto as you move through this book. The AI Efficiency Gap is not proof that AI failed. It's proof that the business hasn't yet installed the measurement discipline needed to know whether AI is working. That's a solvable problem. And solving it starts with understanding exactly what the gap looks like from the inside.

To make this concrete, consider a hypothetical scenario that captures what many operators are experiencing. Return for a moment to the 75-person construction firm carried forward from the earlier Volumes. The CFO there adopted an AI tool for invoice processing about a year ago. Usage is high. The accounts payable team says the tool is helpful. But when the CFO sat down to prepare a quarterly operations review, he realized he couldn't tell the board whether invoice cycle time had actually dropped, whether error rates had improved, or whether the tool had reduced labor hours in any measurable way. He had a license cost, a usage report, and a general sense that things were better. He didn't have a number. That's the AI Efficiency Gap in practice, and it's far more common than most leadership teams are comfortable admitting.

What that CFO did next is the operating beat the rest of this Volume builds out. He did not change the AI tool. He did not call the vendor. He did not start a new initiative. He pulled the per-use-case governance dossier that the controller had produced through the AI Risk & Governance Review™ in the prior quarter. The dossier listed three corrective actions against the invoice processing workflow: ownership of AI output review was undefined, the workflow had no documented baseline for cycle time or error rate, and a vendor risk inventory item was open against the AI provider for its data handling. He took those three findings off the shelf and assigned them to a named owner with a thirty-day completion date. Then he asked his accounts payable lead to spend two weeks counting how long an invoice took to clear the workflow from the moment it arrived to the moment it posted to the general ledger. That two-week count was the missing baseline. The corrective actions were the missing controls. Once both existed, the AI Efficiency Gap was no longer a gap. It was a workflow with a measurement, a set of fixes in motion, and a documented record that any board member could review without anyone in the room having to apologize for the answer.

Why Standard KPIs Miss the Gap Entirely

Traditional KPIs are useful. Revenue per employee, output volume, task completion rates, utilization, and cost per unit are all legitimate measures of business performance. The problem isn't that they're wrong. The problem is that they weren't built to show what AI is doing inside the workflow before the final business result appears. They measure outcomes after the operational cost has already been absorbed. By (Graham et al., 2015) the time a standard KPI reflects a problem, the damage is already done, and it's already been paid for.

AI creates a specific kind of hidden cost that standard KPIs miss almost entirely (Cokins, 2004) .

When AI is inserted into a workflow without proper mapping or measurement, it often moves work rather than removing it. (Acemoglu & Restrepo, 2018) A task that used to take one person two hours might now take thirty minutes with AI, but then require forty-five minutes of review, fifteen minutes of correction, and a second approval step that didn't exist before. The output arrives faster. The KPI looks better. But the total labor cost of producing that output hasn't dropped at all. In some cases it's higher, because now two people are involved instead of one and the process has more steps than it did before AI was introduced.

Standard KPIs won't catch that pattern. Revenue per employee won't show it. Output volume won't show it. Task completion rates won't show it. The numbers at the end of the reporting period will look stable or even slightly positive while AI is quietly creating more review burden, more rework, inconsistent quality, duplicated effort, and management overhead beneath the surface.

To find the AI Efficiency Gap, leaders need measures that live inside the workflow, not at the end of the financial reporting period. Cycle time per task. Review touches per output. Correction rate. Rework hours per week. Handoff delays between stages. Output consistency across team members using the same AI tool. Capacity created without adding headcount. These are the measures that show what AI is actually doing to the process, not just what the process is producing at the end.

The reason this distinction matters so much is that AI performance is a workflow-level event (Devarajan et al., 2024) . The efficiency gain or loss happens inside the process, step by step, handoff by handoff. By the time it shows up in a quarterly KPI, it's already been averaged out, absorbed into overhead, or masked by other variables. If leadership only looks at the final number, they'll miss the AI Efficiency Gap every single time, even when it's costing the business real money every single week.

This is also why the question "are our standard metrics moving?" is the wrong question to ask about AI performance. The right question is "what is happening inside the workflow at each step where AI is involved, and is that step getting faster, cheaper, more consistent, or more error-prone?" That's a workflow-level question. And it requires workflow-level measurement to answer it honestly.

The Connection Between Volumes I Through III and Volume IV

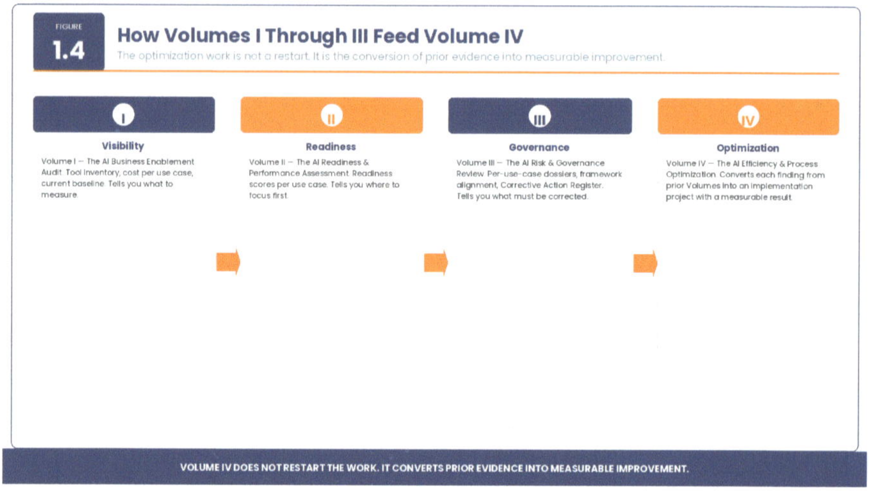

Volume IV doesn't restart the AI conversation. It receives the operating evidence that the first three volumes created and converts that evidence into measurable improvement. That distinction matters because it means the work already done inside your business isn't background context. It's the actual input that makes optimization possible.

Volume I created visibility. It identified which AI tools were active, what each tool cost, what risks each use case introduced, and what the business's current performance baseline looked like across the workflows AI was touching. That inventory is the foundation everything else is built on. Without it, optimization is guesswork. With it, you know exactly where to look and what to measure.

Volume II tested readiness. It evaluated which AI use cases were ready to expand, which ones needed refinement, and which ones should pause. Those readiness decisions aren't just historical records. They're operating signals. A use case that scored low on readiness in Volume II is a use case that probably still has an efficiency problem today, and that problem is exactly where this volume needs to focus first.

Volume III documented governance. It produced the per-use-case governance dossiers, the AI Governance Maturity Scale™ score, the entries in the AI Corrective Action Register™, the findings from the AI Data Exposure Model™, the framework-alignment gaps captured in the AI Governance Framework Crosswalk™ against ISO/IEC 42001, the NIST AI RMF, and the EU AI Act, and the approved target state and migration plan that Steps 5 and 6 of the 6-Step Review Process™ produced for each use case. Those documents are not background paperwork. They are the named inputs into the optimization work this Volume installs. Volume III identifies the ownership gaps, control weaknesses, corrective actions, vendor exposure, data handling issues, and framework gaps. This Volume converts each of them into an implementation project, a workflow redesign, a technical safeguard, a contract remediation effort, or a measurable business improvement. That is the relationship between governance and optimization, and it is the operating logic that holds the AI Operating System™ together.

The way these three Volumes feed into Volume IV is direct and practical. The tool inventory from Volume I tells you what to measure. The readiness scores from Volume II tell you where to focus first. The governance dossiers, the AI Governance Maturity Scale™ score, the AI Corrective Action Register™ entries, the AI Data Exposure Model™ findings, the AI Governance Framework Crosswalk™ alignment gaps, and the approved target states and migration plans from Step 5 and Step 6 of the 6-Step Review Process™ tell you what must be corrected before the efficiency numbers can be trusted and, more importantly, what specific implementation work this Volume is responsible for delivering. Together, they give Volume IV a starting point that most optimization efforts never have: a documented, structured picture of where AI is operating, what the known problems already are, and what an approved future state has already been signed off as the destination.

If you've completed the earlier volumes, you're not starting from scratch. You're starting from a position of documented knowledge, and that's a significant advantage. The work ahead isn't about discovering where AI is operating inside your business. It's about measuring what that operation is actually producing and converting those measurements into decisions that improve performance. That's a much faster and more precise process than building from zero.

For leaders who are coming to this volume without the earlier three, the same logic applies. You'll need to build some of that foundational visibility as you move through the early chapters, and this book will guide you through that. But the underlying principle is the same: optimization requires a documented starting point, a mapped workflow, and a clear measurement system. Those three things are what the first three volumes provide, and what the early chapters of this volume will help you establish if you don't already have them.

Why This Volume Does Not Repeat Volume II's Workflow Readiness Work

A reader who completed Volume II will recognize the workflow language in this Volume and reasonably ask what is different. Volume II tested workflow readiness. This Volume tests workflow performance. The distinction matters operationally because the two questions look adjacent and are not. They ask different things about the same workflows, they produce different decisions, and conflating them weakens both pieces of work.

Volume II asked whether a workflow met the readiness conditions before AI was introduced or expanded. Was the process stable enough. Was the data reliable enough. Was the team prepared to absorb the change. Were the inputs and outputs measurable. Was ownership clear. Was the risk profile acceptable. The output of Volume II was a decision: expand, refine, or pause. That decision is a gating function. It tells the business whether to move forward, hold position, or step back before AI deepens its place inside the workflow.

This Volume picks up after that decision has been made. The workflow has been expanded, refined, or paused. AI is operating inside it. The readiness conditions exist or have been built. What this Volume asks is different. It asks what the workflow is now producing that it was not producing before AI was introduced. It asks whether cycle time dropped, whether capacity opened up, whether error rates declined, whether rework fell, and whether the business has the documentation to defend any of those answers. The output is not a gating decision. It is a measurement, a financial result, and a sequenced improvement plan that closes the distance between what AI promised and what it has actually delivered.

The clean way to hold the distinction is this. Volume II asks whether this workflow is ready for AI to operate inside it. Volume IV asks, now that AI is operating inside this workflow, what is it producing in numbers the business can defend. Both questions are valid. Neither one answers the other. A business that runs only Volume II readiness work ends up with workflows that were approved for AI but never measured. A business that runs only Volume IV performance work ends up measuring workflows where the readiness conditions were never installed, which means the numbers it produces will be unreliable. The two Volumes are sequential by design. Run them in order. Use them for what each was built to do.

The Central Thesis: Governance Identifies What Must Change. Optimization Is the Mechanism That Changes It

Governance is essential. But governance alone doesn't move the business. (Dignam, 2019) A governance review can reveal unclear ownership, weak controls, poor documentation, vendor exposure, data handling issues, and a list of corrective actions that runs several pages long. That's valuable work. It protects the business. It creates accountability. It builds the foundation for defensible AI operations. But it doesn't improve cycle time. It doesn't reduce rework. It doesn't increase capacity or protect margin. Those outcomes require something beyond governance. They require optimization.

Optimization is the mechanism that converts what governance identifies into actual operating change.

Think about what that means in practice. Governance says a workflow has unclear ownership and inconsistent outputs. Optimization maps the workflow, assigns clear ownership, standardizes the AI-assisted steps, measures the output quality before and after, and documents the improvement in a format the CFO can use. Governance identifies the problem. Optimization changes the process. Both are necessary. Neither is sufficient without the other.

The optimization work converts the findings from Volume III into five named output categories, and those five categories cover every kind of operating change a governance review can ask for. The first is implementation projects, where a target-state control or a corrective action becomes a sequenced piece of work with a named owner and a date. The second is workflow redesigns, where a process that governance flagged as unclear, inconsistent, or poorly handed off is rebuilt around the way the work actually moves. The third is technical safeguards, where a data exposure finding or an access control gap from the AI Data Exposure Model™ is closed by a specific technical change. The fourth is contract remediation efforts, where a vendor risk inventory finding or a third-party governance gap becomes a renegotiation, an addendum, or a replacement. The fifth is measurable business improvements, where the work is judged not by the document it produces but by the change it creates in cycle time, capacity, error rate, rework, margin, or ROI. Every finding from Volume III lands in one of these five categories. That is what makes the handoff between governance and optimization operational rather than rhetorical.

This is the central thesis of this book, and it's worth being direct about why it matters. The AI Operating System that Volumes I through III helped you build is incomplete without Volume IV. A business that can see its AI use cases, assess their readiness, and defend its governance decisions has done serious, important work. But if it can't prove that AI is improving the way the business performs, the AI Operating System is still producing documentation rather than operational return. Documentation is not the goal. Measurable improvement is the goal. Documentation is just the evidence that improvement is real and defensible.

Optimization without governance is dangerous. It chases speed without control. (Dafoe, 2018) It improves cycle time in workflows that have unclear ownership and weak quality checks, which means the faster output may be less reliable, less consistent, and less defensible than the slower output it replaced. That's not an improvement. That's an accelerated version of the same problem.

Governance without optimization is incomplete. It creates a record of what needs to change without producing the operating results that justify the AI investment. A business that governs its AI well but can't prove what AI has improved is still sitting in the same uncomfortable position when the CFO asks for the numbers. The governance record shows that the business is responsible. The optimization results show that the business is performing. Leadership needs both to answer the question that eventually arrives in every boardroom.

The AI Efficiency Gap Across Time: Six Months, Twelve Months, and Eighteen Months

The gap does not stay the same shape over time. It presents differently at different durations, and reading the duration correctly changes which corrective actions will work. A gap at six months looks like enthusiasm fading into uncertainty. A gap at twelve months looks like budget questions and quiet team frustration. A gap at eighteen months looks like permanent operating drag that the business has stopped trying to fix. The work in this Volume is most effective when the gap is six to twelve months old, becomes harder at eighteen months, and becomes a multi-quarter recovery project beyond that.

At six months, the gap is fresh enough that the original implementation team is still in place, the workflows have not yet been frozen into permanent patterns, and the measurement infrastructure can be added without unwinding too much that has already been built. The team usually wants to close the gap. The CFO has not yet decided to defund the AI investment. The board has not yet started asking why. This is the cheapest moment to do the work in this Volume, and it is also the moment when most leadership teams have not yet recognized the gap as a structural issue. The early-gap diagnostic is the easiest to run and the hardest to convince the team to take seriously.

At twelve months, the gap has settled into a pattern. The team has worked out informal ways to manage around it. The reports being generated may have started to describe activity as performance, because the team has been trained by quarterly review cycles to report what the dashboard captures. The CFO is starting to ask harder questions during budget reviews. The board has begun making references to AI value that nobody can quite measure. This is the most common state in which leadership teams open Volume IV, and the work proceeds at a steady pace from here, taking two to three quarters to fully close the gap. The methodology in this Volume was built for this duration.

At eighteen months, the gap has become structural. Compensation conversations may have referenced AI productivity that was never measured. Departments may have hired or not hired based on assumed AI capacity that did not materialize. Customer commitments may rest on AI performance that was reported but never verified. The remediation work at this duration is heavier because there are more dependent decisions to unwind, and the financial honesty required in the first measurement cycle is correspondingly harder. The work still produces results, but the first Brief tends to be harder to deliver because it contains more reversed claims than a six-month or twelve-month version would.

The general rule worth holding onto is that the gap compounds. Every quarter the gap stays open, the cost of closing it grows (Chui & Francisco, 2017) , the number of dependent decisions resting on the gap grows, and the political difficulty of delivering the first honest measurement grows. Leadership teams that recognize this and move at six or twelve months close the gap with less disruption than those that wait. The question to ask is not whether the business has the time to do the work in this Volume. It is whether the business has the time to keep deferring it.

How to Run the First AI Efficiency Gap Diagnostic Without Disrupting Operations

The first diagnostic does not require a project plan, a consulting engagement, or a multi-week initiative. It requires a focused two-hour leadership session built around three specific questions, answered honestly, with the resulting documentation captured in writing before the session ends. Most leadership teams find that two hours produces enough evidence to confirm whether the gap exists, where it is concentrated, and what the next quarter of work should focus on. The diagnostic does not slow the team because it does not involve the team. It is leadership work done with leadership data.

The three questions are these. First, for each active AI use case, what does the business currently report about its performance, and where do those reports come from. The leadership team writes down the actual reports being produced (usage rates, output volumes, adoption percentages, savings claims) and the source of each one (vendor dashboard, internal tracking spreadsheet, team self-report). The exercise alone surfaces the pattern. If every reported metric traces to activity counts, the gap is present. If at least one metric traces to a documented baseline comparison, the gap may be partial.

Second, for each active use case, what did the workflow look like before AI was introduced and what does it look like now, mapped step by step in two columns. The exercise does not require detail. Five to seven steps per workflow is enough. The output is two columns side by side: pre-AI workflow steps and current workflow steps. If the workflow was never mapped before AI was introduced, that fact is recorded. If the current workflow has steps the team has been doing for months that nobody had noticed before, that fact is recorded too. The two-column comparison is the most reliable two-minute diagnostic for whether a workflow has been mapped or assumed.

Third, for each active use case, what would a board member learn from a current operational review that the team itself does not already know. This question tests whether the existing reporting tells a story or hides one. A use case where the answer is the same in both directions has reasonably aligned reporting. A use case where the team knows things about the workflow that the board reports would never reveal has hidden performance information. A use case where the board reports tell a story the team would describe differently has misaligned reporting. All three states are useful to surface, and the question takes about ten minutes per use case to answer honestly.

The output of the two-hour diagnostic is a one-page summary listing each active AI use case, its current reporting source, its workflow-mapping status, and its reporting alignment status. That single page is the starting position for the rest of the work in this Volume. The team does not need to be involved until the corrective work begins. Most leadership teams who run the diagnostic come out of the session with a sharper view of where the gap lives in their business and a clearer sense of which two or three workflows would benefit most from the optimization work that follows.

What Three Engagements Look Like When the Gap Has Been Closed

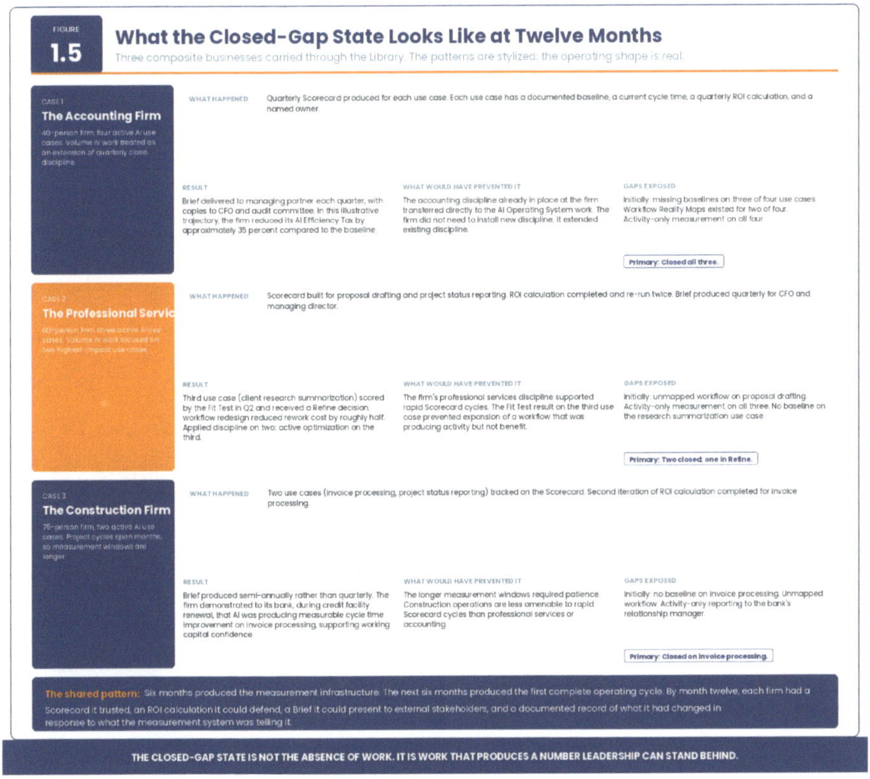

The clearest way to understand what the closed-gap state looks like is to walk through the three composite businesses carried through this Library and describe what each one's operating posture looks like twelve months after starting the Volume IV work. The composites are stylized but the patterns are drawn from actual engagement experience.

The 40-person accounting firm twelve months in has built the most complete operating record of the three. Accountants are trained on documentation and measurement, and the firm's leadership treated the Volume IV work as an extension of its quarterly close discipline. The firm now produces a Scorecard each quarter for the four AI use cases in its operations: client document review, draft tax return preparation, audit workpaper drafting, and client communication drafting. Each use case has a documented cycle time baseline, a current cycle time, a quarterly ROI calculation, and a named owner. The Executive AI Efficiency Brief goes to the managing partner each quarter, with copies to the CFO and the audit committee. In the illustrative trajectory shown here, the firm has reduced its AI Efficiency Tax by approximately 35 percent compared to the baseline reading taken at the start of the work; actual results in any specific engagement depend on the workflows targeted, the starting baseline, the leadership team's commitment to the measurement and standardization work described in this Volume, and operating factors specific to each business. The closed-gap posture here looks like accounting discipline applied to a new operational area.

The 60-person professional services firm twelve months in shows a different shape. The firm runs three AI use cases (proposal drafting, project status reporting, and client research summarization) and decided to focus the Volume IV work on the two highest-impact ones (proposal drafting and project status reporting). The Scorecard exists for both. The ROI calculation has been completed and re-run twice. The Brief is produced quarterly and read by the firm's CFO and managing director. The third use case (client research summarization) was scored by the AI Process Fit Test in the second quarter and received a refine decision, which produced a workflow redesign that reduced the rework cost by roughly half. The closed-gap posture here looks like applied discipline on two workflows and active optimization on the third.

The 75-person construction firm twelve months in shows the longest road. Construction operations are less amenable to the rapid Scorecard cycle than professional services or accounting, because the underlying projects span months and the cycle time measurement requires longer reading windows. The firm tracks two AI use cases (invoice processing and project status reporting) on the Scorecard and is on the second iteration of the ROI calculation for invoice processing. The Brief is produced semi-annually rather than quarterly because the underlying data takes longer to settle. The firm has been able to demonstrate to its bank, during a credit facility renewal, that AI is producing measurable cycle time improvement on the invoice processing workflow, which directly supported the bank's confidence in the firm's working capital management. The closed-gap posture here looks like measured progress on the workflows where measurement is feasible and patience on the workflows where it is not.

Across all three engagements, the common pattern is this. The first six months produced the measurement infrastructure. The next six months produced the first complete operating cycle. By twelve months the firms had a Scorecard they trusted, an ROI calculation they could defend, a Brief they could present to external stakeholders, and a documented record of what they had changed in response to what the measurement system was telling them. The closed-gap state is not the absence of work. It is the presence of work that produces a number the leadership team can stand behind.

The reader's job, starting with this chapter and running through every chapter that follows, is to close the loop. Take what the business already knows from the earlier volumes. Take what it already scored in the readiness assessment. Take what it already documented in the governance review. Then convert that existing evidence into measurable improvement that a CFO, a board member, a lender, or an acquirer can trust. That's what optimization means in the context of this book. Not speed for its own sake. Not efficiency as a talking point. Measurable operational improvement that shows up in cycle time, capacity, error rates, labor savings, margin, and ROI, and that can be defended with numbers when someone in the room asks to see them.

Putting It Into Practice

Before moving into the next chapter, there's one task that will make everything that follows more useful. You need to name the gap inside your own business. Not in general terms. Specifically.

Start by identifying the three AI-supported workflows in your business where AI activity is highest. These are the workflows where employees are using AI tools most frequently, where output volume is highest, or where leadership has the most enthusiasm about AI's contribution. Write them down. These three workflows are where the AI Efficiency Gap is most likely hiding, because high activity without measurement is exactly where the gap lives.

For each of those three workflows, run through four questions. First, what was the baseline? What did cycle time, error rate, rework volume, or cost per output look like before AI was introduced? If you don't have a documented baseline, write that down. The absence of a baseline is itself a finding, and it's the first thing that needs to be corrected. Second, what measurably changed? Not what the team believes changed. What can the business actually show changed, with a number attached to it? Third, what hidden cost appeared? Has review burden increased? Has rework grown? Have new approval steps been added? Has management time gone up? Fourth, what financial result can the business currently defend? Not estimate. Defend, with documentation, to a CFO or board member who is skeptical and asking hard questions.

Most leadership teams who run through those four questions honestly will find that they can answer the first question partially, the second question vaguely, the third question not at all, and the fourth question with something that sounds more like a story than a number. That's the gap. That's exactly what this book is built to close.

The final move before Chapter 2 is a leadership conversation. Bring your findings from the four questions above into your next operating meeting and say this clearly to your team: the business is no longer measuring whether AI is being used. It's measuring what AI has moved. That shift in language signals a shift in accountability. It tells the team that usage reports and adoption rates are no longer the finish line. Measurable improvement is the finish line. (Babar et al., 2025) And it sets the expectation that every AI-supported workflow in the business will eventually need to produce a number that can be defended, not just an activity report that can be shared.

That conversation is where optimization begins. Not with a tool. Not with a framework. With a clear statement from leadership that the standard has changed, and that AI will now be held to the same accountability standard as every other operating investment in the business.

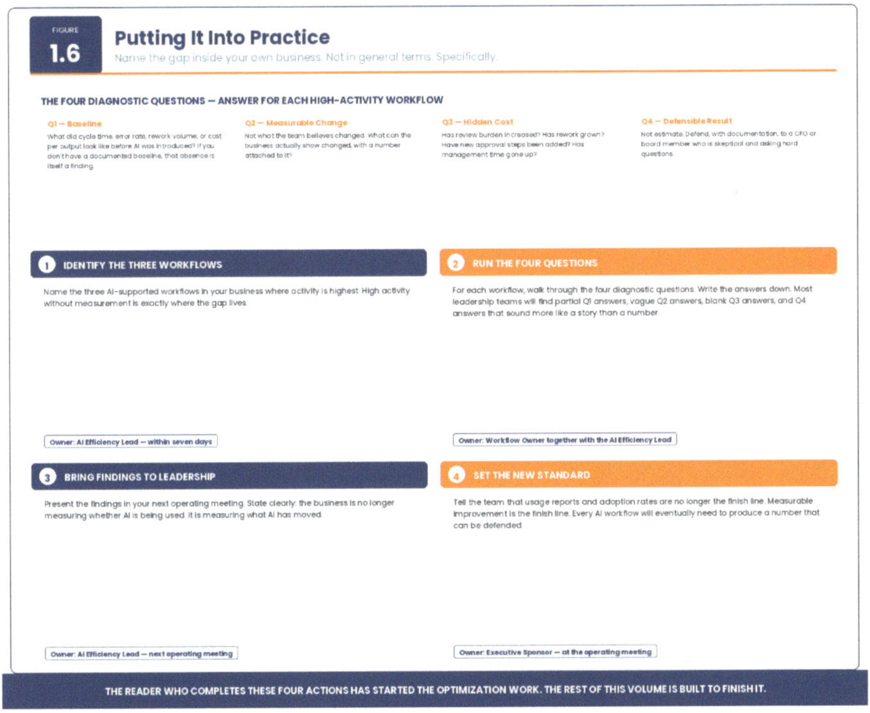

Chapter 2: Phantom Productivity

When AI Looks Like It's Working and Isn't

CHAPTER 2 OUTCOMES

At the end of this chapter, you will be able to:

1. Define Phantom Productivity and distinguish it from real efficiency gains and from legitimate setup friction.
2. Recognize the three conditions inside a workflow that create Phantom Productivity, and the four signals that confirm it.
3. Identify where Phantom Productivity first surfaces in the P&L and what to look for on the income statement.
4. Run the upstream-versus-downstream measurement that converts a suspicion of Phantom Productivity into evidence.
5. Address Phantom Productivity findings with the leadership team without blaming the people who experienced the AI as helpful.

PHANTOM PRODUCTIVITY IS NOT A TOOL PROBLEM. IT IS A WORKFLOW PLACEMENT AND MEASUREMENT PROBLEM.

The Problem This Chapter Solves: Your Team Looks More Productive But the Business Hasn't Moved

There's a specific kind of frustration that shows up in leadership teams about six to twelve months after serious AI adoption. The tools are running. The team is using them. Output volume is up. The dashboards look healthy. And yet, when a CFO or board member asks what AI has actually improved in the business, nobody has a clean answer. The workflow doesn't feel faster. The margin hasn't moved. The rework hasn't dropped. Something looks productive, but the business itself hasn't changed in any way that can be measured or defended.

That condition is what Chapter 1 named the AI Efficiency Gap. This chapter names the most common specific pattern through which the Gap appears in practice: Phantom Productivity. It has a specific relationship to the operational friction work in Volume II. Volume II Chapter 7 introduced the AI Friction Diagnostic and identified three friction types that appear when AI runs inside a business: correction friction, confidence friction, and bypass friction. That diagnostic was built to surface friction during readiness assessment, before broader scaling decisions. Phantom Productivity is what those same frictions look like when they have been allowed to persist past the readiness phase, settled into the operating rhythm, and started producing measurable financial drag. Volume II asks whether the friction exists. This Volume measures what the friction has cost the business that has been absorbing it for the last six to twelve months. The friction Volume II identified is the same friction this Volume puts a dollar figure on; the angle is what changes.

Phantom Productivity is what happens when AI increases visible output without improving the underlying business result. Cycle times stay flat. Rework rises quietly. Review burden grows. Quality becomes inconsistent across the team. Margins don't move. And because the outputs look polished and the activity looks real, nobody in the organization connects those problems back to the AI tools that created them. The business celebrates the output at the exact moment the process is becoming more expensive to run.

This is rarely an employee failure. It's rarely a technology failure either. It's almost always a measurement failure. The business started rewarding activity before it tested whether that activity produced a better downstream result. And once that pattern takes hold, it compounds quickly. More AI usage gets approved. More workflows get touched. More outputs get generated. And the hidden cost of reviewing, correcting, and managing all of that output keeps growing in the background, invisible on every dashboard and absorbed silently into labor hours that nobody is tracking back to the source.

What follows shows how Phantom Productivity forms, what it looks like from inside an active AI workflow, and how to detect it before it compounds further. For any leader who has looked at AI usage reports and still wondered why the business doesn't feel faster, cleaner, or more profitable, the diagnostic lens in this chapter shows exactly where the gap is hiding.

What Phantom Productivity Actually Looks Like Inside a Business

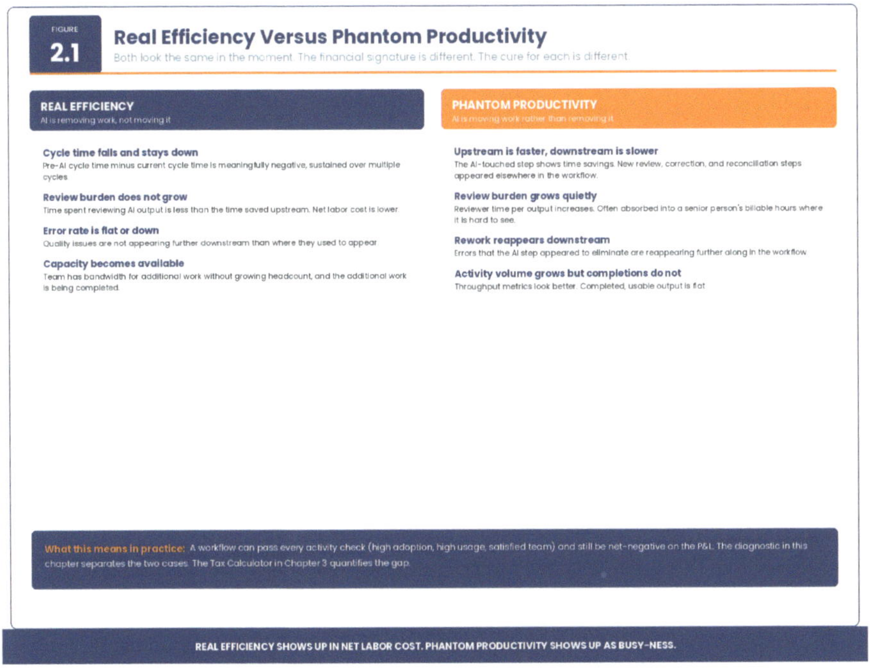

Phantom Productivity doesn't announce itself. It looks positive right up until the moment someone follows the work downstream and starts asking what happened to it after AI created it.

The most common pattern shows up in content and communication workflows. AI generates a draft. The draft looks polished. It has structure, complete sentences, and a professional tone. But when the reviewer opens it, they spend twenty minutes fixing the facts, adjusting the tone, removing assumptions the AI made that don't apply to this specific client, and adding context that was missing entirely. The draft arrived faster. The total time to produce a finished, usable output didn't drop at all. In many cases it went up, because now a second person is involved in a review step that didn't exist before the AI tool was introduced.

That's Phantom Productivity in its most common form. More upstream output, same downstream result, more people involved in getting there.

The pattern also shows up in reporting and data workflows. AI produces more summaries, more reports, more compiled data sets. The volume of information flowing through the organization increases. But approvals don't get faster. Decisions don't get sharper. Internal teams still wait just as long for the information they need to act on. The AI created more material without creating more clarity, and the people receiving that material now spend more time sorting through it to find what's actually useful.

The 75-person construction firm carried through this Library encountered this exact pattern. Volume II Chapter 6 describes the firm's discovery that drafting time dropped from three hours to under six minutes after AI was introduced, while downstream correction time expanded from thirty minutes to four hours, producing a net delivery cost that was higher than the pre-AI workflow. The Volume II analysis identified the friction. The Volume IV question is what the firm did about it once the friction was identified. That answer requires the diagnostic and corrective discipline the rest of this chapter installs.

Other signals to watch for include fewer blank pages but more revision cycles, higher task completion rates on the front end with longer approval queues on the back end, and employees who report feeling more productive while managers quietly spend more time inspecting and correcting the work coming through. Phantom Productivity often hides behind good-looking work, which is what makes it more dangerous than obvious failure. When something fails visibly, the organization responds. When something looks successful while quietly becoming more expensive, the organization keeps investing in the problem.

The Three Conditions That Create Phantom Productivity

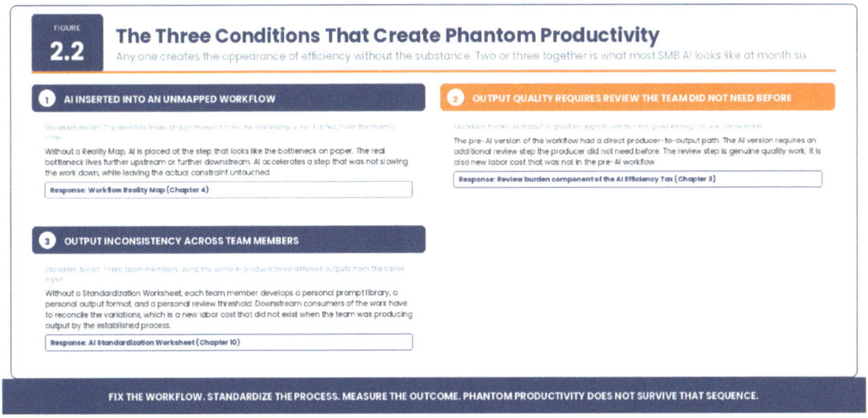

Phantom Productivity doesn't appear randomly. It forms when specific operating conditions exist around the AI-supported workflow, and the three conditions map directly to three of the six readiness conditions Volume II's AI Readiness & Performance Assessment evaluates. Where Volume II's assessment scores these conditions before AI is expanded, Phantom Productivity is what happens when these conditions remain weak after AI has already been expanded. The Volume II treatment is the readiness check; the Volume IV treatment is the recovery work.

The first condition is an unmapped or unstandardized workflow. When AI gets dropped into a process that the business hasn't clearly defined, the tool accelerates whatever the process already does, including its inefficiencies, ambiguities, and failure points. If the workflow had unclear handoffs before AI, it now has faster unclear handoffs. If the workflow had inconsistent inputs before AI, it now produces inconsistent outputs at higher volume. AI doesn't fix an undefined process. It runs faster inside it, which means the problems compound more quickly and at greater scale (Mueller & Massaron, 2021) than they did before the tool was introduced.

The second condition is unmeasured output quality. AI work often looks polished on the surface. Formatting is clean. Sentences are complete. The output has the visual appearance of professional work. But polish is not accuracy. Polish is not consistency. Polish is not relevance to the specific client, context, or decision at hand. When no one is tracking accuracy rates, consistency across team members, review time, client impact, or correction frequency, the organization has no way to know whether the polished output is actually better than what it replaced. It just knows the output arrived faster and looks good. That's not a measurement. That's an impression, and impressions don't hold up when a CFO asks for the numbers.

The third condition is rewarding volume over value. This one is cultural, and it's the hardest to correct because it's often built into how leadership talks about AI performance. When teams are praised for producing more drafts, more summaries, more reports, more responses, the incentive structure points toward output volume. Employees respond to that incentive rationally. They produce more. The activity metrics go up. The usage reports look strong. But if the downstream result hasn't improved, all that additional output is generating cost without generating value. The business is paying for AI tools, paying for the labor that runs those tools, and paying again for the review and correction work that cleans up the output before it can actually be used. Volume without value is an expensive loop, and it runs continuously until leadership changes what it measures and what it rewards.

When all three conditions are present together, Phantom Productivity is almost guaranteed. The diagnostic question for any AI-supported workflow is straightforward: did AI increase output but leave cycle time, rework, review burden, and cost per completed unit unchanged or worse? If the answer is yes, the business is measuring motion instead of improvement. And motion without improvement is exactly what Phantom Productivity produces.

Distinguishing Phantom Productivity from Legitimate Setup Friction

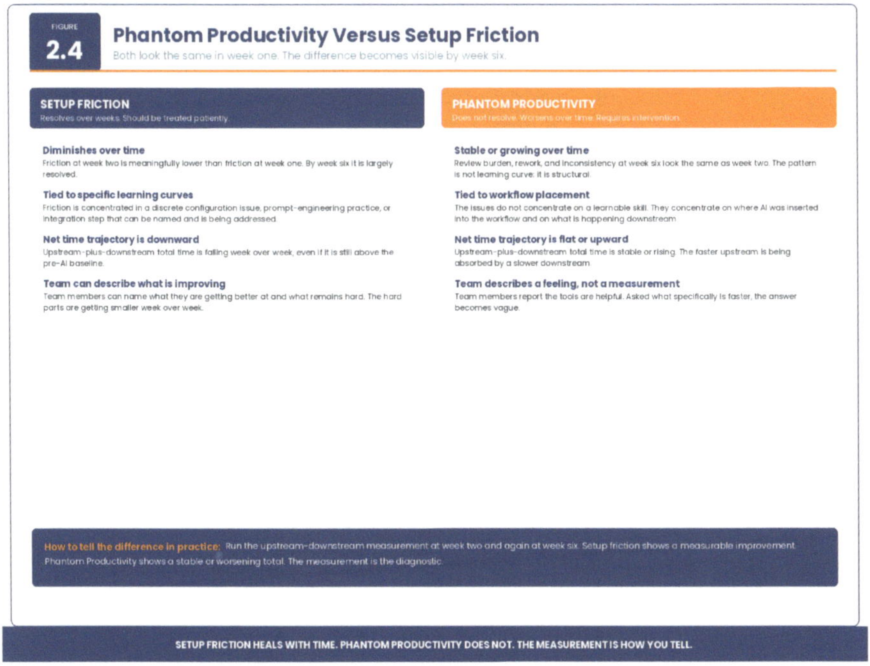

Not every AI workflow that takes longer than expected in its early months is running Phantom Productivity. New tools require setup. New prompting standards require practice. New review conventions require time to settle. A workflow that runs slower in months one through three after AI was introduced may be in setup friction, which is normal and time-bounded, or in Phantom Productivity, which is structural and worsens with scale. The two patterns look identical on a usage report. They are different on the operational evidence, and conflating them produces the wrong response: walking back a use case that was about to clear setup friction, or doubling down on a use case that is structurally generating cost.

The cleanest distinction runs through four observable signals. Setup friction shows declining review burden month over month as the team learns. Phantom

Productivity shows flat or rising review burden. Setup friction shows rework concentrating in a known list of issues that the team is actively fixing. Phantom Productivity shows rework distributed across the workflow, with no clear pattern of resolution. Setup friction has an explicit end date in the project plan, after which the team expects normal operation. Phantom Productivity has no end date because nobody has identified it as a problem yet. Setup friction is named, owned, and measured. Phantom Productivity is unnamed, unowned, and absorbed.

The practical test is simple. Run the four signals against any AI workflow that is consuming more labor today than it was projected to consume when AI was introduced. If the signals point to setup friction, give the workflow another sixty days and re-measure. If the signals point to Phantom Productivity, the workflow needs the diagnostic work in the rest of this chapter applied immediately, because the cost is not going to resolve on its own. Most operators, when they run this test honestly, find that one or two of their workflows are in setup friction and one or two are in Phantom Productivity. The temptation is to treat them the same way. The right move is to treat them differently, with the diagnostic effort focused on the second category and the patience preserved for the first.

When the Shadow Tool Is Producing the Result

Phantom Productivity has a complement that is worth naming directly, and it is the first place in the Library where shadow AI appears as a measurement-credibility issue rather than a governance issue. Volume I Chapter 2 introduced shadow AI as a culture and visibility problem, with the discovery work focused on bringing shadow tools into the approved environment. Volume II Chapter 3 treated shadow AI as a data exposure pathway and an internal control failure. Volume III Chapter 4 quantified the breach cost: shadow AI added roughly USD 200,000 to the average breach in the IBM and Ponemon 2025 data. Each of those Volumes addresses what shadow AI costs the business when something goes wrong. This Volume addresses what shadow AI costs the business when nothing goes wrong at all, because the operating result is being produced outside the tool that the measurement system is reporting on.

The pattern looks like this. The AI tool the leadership team approved is producing the appearance of efficiency, while a different tool, often a personal subscription the team adopted on its own, is producing the actual result. The approved tool runs the usage report. The shadow tool runs the workflow. The output volume the dashboard shows traces to the approved tool. The quality of the work that reaches the client traces to the shadow tool, which leadership does not see, does not measure, and does not pay for through the corporate plan. The consequence is that every measurement downstream of the approved tool, including the AI Efficiency Scorecard in Chapter 7 and the AI ROI Formula in Chapter 8, is being calculated against the wrong source. The number the business is reporting on AI performance is not the number the business is actually producing. That is a measurement-system failure, and it is what makes shadow tools a Volume IV concern even after Volumes I, II, and III have done their work.

This pattern is more common than most leadership teams expect. The detection method is the same downstream measurement described in this chapter, but the question changes. Once the downstream measurement shows a result that the upstream activity cannot explain, the next question is which tool actually produced the result. The answer is usually a brief, candid conversation with the people doing the work. The conversation requires the same framing the Volume III governance review used. The business is not investigating misconduct. It is investigating which tools are actually producing the operating result, so that the AI Operating System™ reflects the real workflow rather than the documented one.

Once the shadow tool is named, leadership has a decision to make. Bring the shadow tool inside the approved environment with the governance, security, and data handling controls that the AI Governance Framework Crosswalk™ requires. Or stop using it and rebuild the approved tool by upgrading its prompting and review structure to actually produce the result the team needs. Both are valid moves. The wrong move is to keep paying for the approved tool while the team continues to produce the real work outside the system the business is governing.

Phantom Productivity Is Not a Tool Problem

Phantom Productivity is rarely a tool problem. It is almost always a workflow problem, a measurement problem, or an incentive problem. The tool produced output. The workflow accepted output that was not yet ready for downstream use. The measurement system rewarded the upstream count without checking the downstream result. Replacing the tool while leaving the workflow, the measurement system, and the incentive structure intact produces the same pattern with a new vendor, and the business absorbs the switching cost for no operating gain. The corrective move is to fix what surrounds the tool before evaluating whether the tool itself needs to change. Most Phantom Productivity patterns resolve when the workflow is mapped accurately, the output standards are written, the review checkpoint is moved, and the measurement system starts tracking what the business actually needs rather than what the tool happens to produce.

The Three Conversations Phantom Productivity Requires

Once Phantom Productivity has been detected and the cause has been separated from the tool, the leadership team needs to hold three specific conversations before any operational change moves forward. Each conversation has a different audience, a different purpose, and a different outcome. Skipping any of them produces a fix that does not hold past the first quarter.

The first conversation is with the team doing the work. The framing is direct. The business has discovered that the AI workflow is producing more output than the business is gaining from. That is not the team responsibility. It is a process and measurement gap that leadership did not catch early enough. The purpose of this conversation is to surface the specific friction points, shadow steps, and rework patterns the team has been absorbing quietly. The outcome is a documented list of what is actually happening inside the workflow, which becomes the input to the redesign work.

The second conversation is with the people downstream of the workflow. This includes account managers, client services, finance, or whoever receives the AI output and acts on it next. The framing is also direct. We want to know what happens to the AI output after it leaves the producing team. The purpose is to confirm whether the downstream cost is review, correction, rework, or something else, and to quantify it in hours per week. The outcome is a second documented list, this one from the downstream perspective, which usually does not match the upstream list and which together produces the complete picture.

The third conversation is with the executive team. The framing here is different. The other two conversations were about gathering evidence. This one is about acting on it. The purpose is to decide what changes inside the workflow, who owns the change, and what the timeline is for measuring whether the change worked. The outcome is a documented decision that goes into the AI Operating System™ record, so that when the next quarterly review happens, the team has a baseline and a set of corrective actions against which performance can be measured.

Most leadership teams try to compress these three conversations into one and lose two-thirds of the information that would have surfaced if each conversation had been held separately. The three are not redundant. Each one produces information the others do not, and the operational fix only holds when all three pieces of information are on the table at the same time.

To make the diagnostic concrete, return to the 75-person construction firm. Its operations director identified three months of rising correction costs on AI-generated status reports. The output volume tripled. The team felt more productive. But the account managers had quietly stopped using the AI-drafted reports for the firm's largest clients because the corrections required too many hours per report to be worth running through the tool. They were producing those reports by hand instead, while still logging the AI tool as the workflow source in the system. By the time the operations director ran the downstream measurement that this chapter recommends, he found that AI activity in the reporting workflow was three times higher than it had been the prior year, while the actual number of AI-assisted reports that reached a client without manual rewriting had dropped by more than half. The dashboard said the workflow was thriving. The downstream told a different story. He paused expansion of the tool, called the account management team into a workflow review meeting, and used the downstream measurement to build the case for redesigning the review checkpoint and rewriting the prompting guidance before any further expansion was approved. That sequence, starting with the downstream count and ending with a workflow correction grounded in the measurement evidence, is the operating beat this Volume builds out across the next several chapters.

How to Separate Real Efficiency Gains from Phantom Activity

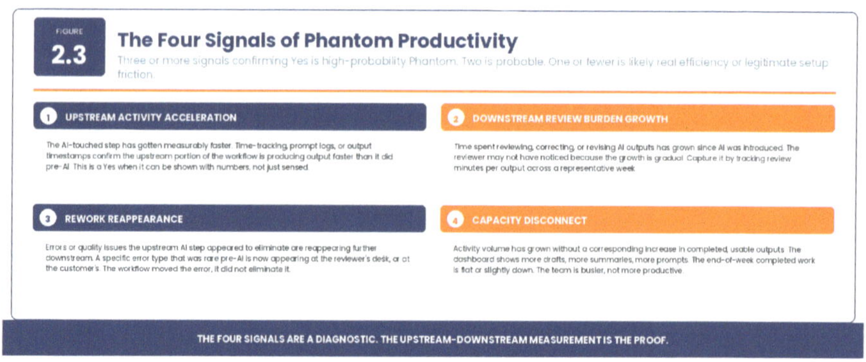

Real efficiency is proven downstream. It's not claimed at the point where AI creates the first draft, the first summary, or the first recommendation. The moment AI generates an output is the beginning of the process, not the end of it. What happens to that output after AI creates it is where real efficiency either appears or doesn't.

The controlling question for every AI-supported workflow is this: did the final business result improve, or did the team simply create more upstream output that still required the same amount of correction, review, approval, and follow-up to reach a usable state? That question cuts through every usage report, every adoption metric, and every activity dashboard. It forces the evaluation to happen at the point where the work actually finishes, not at the point where AI made the work look like it was starting well.

Leading indicators are useful early signals, but they don't prove value on their own (Anderson & Fornell, 2000) . Faster drafts, more completed tasks, higher tool usage, and increased output volume all tell you that AI is active. They don't tell you whether AI is improving the business. A business can have excellent leading indicators and a growing Phantom Productivity problem at the same time, because leading indicators measure what AI is producing, not what the business is gaining from it.

Lagging indicators are where real efficiency shows up. Shorter cycle time from task start to final delivery. Fewer review touches per output before approval. Lower rework hours per week. Better consistency across team members doing the same AI-assisted task. Improved capacity, meaning the team is handling more volume without adding headcount or extending hours. Stronger client response or fewer client corrections. Lower cost per completed unit of work. These are the measures that tell you AI has actually changed how the business performs, not just how busy the team looks while performing it.

To run this test inside your own operation, pick one AI-supported workflow and map it in two versions. The first version is the upstream view: where does AI create an output, how fast does it create it, and what volume does it produce? The second version is the downstream view: how many people touch that output after AI creates it, how many corrections are made before it reaches its final state, how long does the full cycle take from start to finish, and what does it cost in total labor hours to get the work from AI draft to approved deliverable? Compare those two views. If the upstream looks fast but the downstream looks slow, expensive, and correction-heavy, you're looking at Phantom Productivity. The efficiency gain AI appeared to create on the front end was absorbed and then some by the review and correction cost on the back end.

Every AI workflow needs a measurement checkpoint before leadership declares success. The point of proof is not when AI produces something faster. The point of proof is when the business finishes the work better, cheaper, cleaner, or with less strain on the team than it did before AI was introduced. Until that downstream proof exists, the efficiency gain is a claim, not a result.

A practical way to install this checkpoint is to assign one person per workflow the specific job of tracking the downstream numbers for thirty days. Not the AI usage numbers. The downstream numbers: total cycle time, review touches, correction rate, rework hours, and cost per completed output. At the end of thirty days, compare those numbers against whatever baseline exists from before AI was introduced, or against the team's best estimate of pre-AI performance if a formal baseline wasn't captured. The comparison will either confirm that AI is producing real efficiency or reveal that the efficiency is phantom. Either answer is useful. The confirmation tells you which workflows to expand. The revelation tells you which workflows need to be remapped before they absorb any more investment.

How to Address Phantom Productivity Without Blaming the Team

When leadership discovers that AI activity hasn't produced the business improvement they expected, the first instinct is often to look for someone responsible. That instinct is understandable, but it's the wrong move. Almost every employee who contributed to a Phantom Productivity pattern did exactly what they were asked to do. They used the AI tools. They produced more output. They responded to the incentives and expectations leadership set. If the measurement system rewarded volume, they produced volume. If the culture celebrated faster drafts, they delivered faster drafts. The gap between what AI appeared to produce and what the business actually gained is a process and measurement failure, not a people failure.

The conversation that needs to happen is a reset, not a review.

The reset message is direct and non-punitive: the goal was never more AI output. The goal was better business results. Those two things looked similar from a distance, but they're different, and the measurement system didn't catch the difference early enough. Now it will. Going forward, every AI-supported workflow will be measured by what it produces downstream, not by how much it generates upstream. That's the new standard, and it applies to the tools and the processes, not to the people using them.

That framing matters because it keeps the team engaged rather than defensive. Employees who feel blamed for a measurement failure don't fix the process. They protect themselves. (Park et al., 2021) They hide workarounds. They find ways to make their activity numbers look good while the real problem continues underneath. Shadow processes, undocumented shortcuts, and informal workarounds almost always multiply when employees believe that a leadership review is about finding fault (Prescott, 2024) rather than fixing the system. If you want accurate information about where AI is creating review burden, where correction cycles are longest, and where the downstream cost is highest, the team has to trust that sharing that information won't be used against them.

The practical way to build that trust is to make the investigation collaborative from the start. Ask the team directly: where is AI creating more work for you, not less? Where are you spending time cleaning up AI outputs before they're usable? Where does the review step feel heavier than it should? Those questions signal that leadership is trying to fix the process, not evaluate the people. And the answers will surface the Phantom Productivity patterns faster than any top-down audit because the people doing the work already know where the friction is. They just need to know it's safe to say so.

Once the patterns are identified, the correction is operational, not disciplinary. Remap the workflow. Standardize the AI-assisted steps. Set output quality criteria before the review stage rather than discovering quality problems during it. Measure the downstream result and report it alongside the upstream activity so that both numbers are visible to the team at the same time. When employees can see both the output volume and the downstream efficiency result in the same report, they start connecting their own behavior to the business outcome in a way that no policy memo ever produces.

The leadership finish on Phantom Productivity is this: the business doesn't need to produce less AI output. It needs to produce better AI output that travels through the workflow more cleanly, requires less correction, and delivers a measurable result at the end. That's a process improvement conversation. It should feel like one.

Where Phantom Productivity Shows Up First in the Profit and Loss Statement

Phantom Productivity is operationally invisible until it isn't, and when it becomes visible, it usually shows up in the P&L before it shows up anywhere else. Reading the P&L for the early signs of Phantom Productivity is one of the most useful diagnostic skills a CFO can develop, and it is one of the most useful tools a CEO has for testing whether a department's reported AI gains are real. The three signatures that show up first are margin compression in high-AI seasons, capacity that does not translate to revenue, and labor cost that does not scale down with expected AI displacement.

Margin compression in high-AI seasons is the clearest of the three. If a workflow has been told it is faster because of AI, but the margin on the work that flows through that workflow has not improved or has compressed, something is consuming the apparent savings. The team is producing more output, the AI tool is reporting strong usage, and yet the margin on each unit of work is the same as it was before AI or slightly worse. The most common explanation is that the cycle time improvement at the AI step has been absorbed by review burden, rework, or shadow processes downstream. Reading margin by workflow rather than by department surfaces this pattern faster than department-level reporting does.

Capacity that does not translate to revenue is the second signature, and it is often the one a CEO notices before a CFO does. AI was supposed to free up time that the team could redirect to higher-value work, additional client volume, or new revenue-producing activity. The team reports that AI is saving them hours per week. The hours appear on the usage report. But the revenue line has not moved in the direction the freed capacity would suggest, and headcount has not been reallocated to new work. The most common explanation is that the freed capacity was real at the AI step but was reabsorbed by the review and correction work that grew alongside it, leaving the team with the same effective billable or productive hours it had before AI. Reading freed capacity against revenue per employee or revenue per billable hour across the same period is what makes this signature visible. A capacity gain the business cannot trace into a revenue gain is a capacity gain that does not exist outside the report.

Labor cost that does not scale down is the third and most uncomfortable signature. AI was supposed to reduce the labor cost of a workflow. The leadership team approved AI on that assumption. The team did not get smaller. Labor cost did not decline. In some cases it grew, because the team needed more reviewers to handle the AI-assisted output (Checco et al., 2021) . The explanation here is rarely that AI did not work. It is usually that the workflow was redesigned around AI without recognizing that AI shifted the labor location rather than eliminating it. The labor moved from production into review (Chang, 2020) , and the review labor is often higher-skilled and more expensive than the production labor was. Reading total labor cost as a function of output volume across multiple quarters is what surfaces this pattern.

None of these signatures conclusively prove Phantom Productivity. They are leading indicators that warrant investigation. The investigation follows the diagnostic methodology in the body of this chapter. But a leadership team that reads the P&L with these three signatures in mind will catch Phantom Productivity earlier than a leadership team that waits for usage reports to tell them something is wrong, because the usage reports will keep telling them the workflow is thriving until someone runs the downstream measurement that contradicts the report.

How Phantom Productivity Patterns Differ by Function

Phantom Productivity does not look the same in every function. The underlying mechanism is consistent (upstream activity inflated relative to downstream value) but the specific presentations differ by business function, and the corrective actions vary accordingly. The four most common functional presentations are in sales, operations, finance, and customer service.

In sales functions, Phantom Productivity usually presents as a higher volume of AI-drafted proposals, emails, and follow-ups producing the same or lower close rate. The AI tool reports that sales reps are generating more outreach than ever. The volume of activity is impressive. But the conversion rate from outreach to qualified opportunity has flatlined or declined. The explanation is usually that the AI-drafted content has the texture of mass-produced communication, and prospects are responding less. The corrective action is to standardize the AI-assisted outreach to a higher quality bar, even at the cost of volume, and to measure the workflow on conversion rather than on output count.

In operations functions, Phantom Productivity usually presents as more processes documented with AI assistance, faster initial draft cycles, and the same number of operational incidents as before. The team is producing more SOPs, more runbooks, more process documentation. The AI is genuinely faster at draft generation. But operational performance has not improved. The explanation is usually that the AI-drafted documentation is being filed but not actually used, because the team has not changed the underlying operational behavior the documentation was supposed to influence. The corrective action is to measure operational outcomes rather than documentation volume, and to install acknowledgment discipline on the documentation that actually matters.

In finance functions, Phantom Productivity usually presents as more analyses produced faster with the same or worse analytical accuracy. The accounting team is generating more variance analyses, more board memos, more financial models. The AI is producing first drafts in minutes that used to take hours. But errors are appearing in the final outputs at rates higher than before AI, and the senior reviewers are spending more time correcting AI-drafted analyses than they used to spend producing them from scratch. The corrective action is to install structured review checkpoints with documented error-rate measurement, and to track the time the senior reviewer is spending on AI-assisted output versus original output.

In customer service functions, Phantom Productivity usually presents as faster initial response times, higher case volumes, and a customer satisfaction score that has not improved or has declined. The team is handling more cases per hour. The AI is producing first responses in seconds. But customers are escalating more often, or expressing frustration with the responses they received, or coming back to the team a second time with the same issue. The explanation is usually that the AI-drafted responses are technically responsive but operationally unhelpful. The corrective action is to measure first-contact resolution rather than response time, and to track the secondary contact rate as a leading indicator of AI response quality.

The pattern across all four functions is the same. The AI is being measured on what it is good at (volume, speed, immediate output) while the business is being measured on what the output produces (revenue, operational outcomes, customer retention, financial accuracy). The Phantom Productivity gap lives in the space between those two measurements, and the corrective action is always to bring the AI measurement closer to the business measurement.

Reading Phantom Productivity Trends Across Three Quarters

A single quarter of measurement is not enough to identify a Phantom Productivity trend with confidence (Xu, 2014) . A workflow that scores poorly in one quarter may be in setup friction, in a seasonal trough, or in a temporary disruption. A workflow that scores well in one quarter may be benefiting from a temporary tailwind. Three quarters of measurement is the minimum for a Phantom Productivity reading that the leadership team can trust enough to act on, and the way to read those three quarters together produces a different conclusion than reading any one of them in isolation would.

The improving-toward-real-value trend looks like this. Quarter one shows high AI activity, high upstream volume, and low downstream value. Quarter two shows the same upstream volume but slightly higher downstream value. Quarter three shows upstream volume flattening as the team learns to focus AI work where it matters, and downstream value continuing to rise. This trend usually traces to active workflow correction work the team has been doing on its own, often in response to leadership pressure or external feedback. The right response is to formalize what the team has been doing into documented standards and continue the trajectory.

The plateaued-Phantom-Productivity trend looks like this. Quarter one shows high upstream, low downstream. Quarter two shows high upstream, low downstream. Quarter three shows high upstream, low downstream. The workflow is producing more activity than value and has been doing so consistently for nine months. This is the textbook Phantom Productivity state, and the right response is the diagnostic and corrective work described in the rest of this chapter. A workflow in this trend for three quarters will be in this trend for another year if nothing changes, because the team has settled into the pattern and is no longer trying to fix it on its own.

The deteriorating trend looks like this. Quarter one shows balanced upstream and downstream. Quarter two shows higher upstream and slightly lower downstream. Quarter three shows much higher upstream and meaningfully lower downstream. The workflow is moving in the wrong direction. Something specific changed between the quarters, and the diagnostic question is what that something is. Common causes include team turnover that removed the people who knew how to manage the AI output, vendor product updates that changed the AI tool's behavior, and the addition of new use cases that diluted the team's attention on the original one. The corrective action depends on the cause, and the diagnostic to isolate the cause should happen before the corrective action is selected.

Reading the trend requires having the underlying measurements in place across all three quarters. A leadership team that has the AI Efficiency Scorecard running for three quarters has the data to read the trend. A leadership team that does not has to start measurement now, which means the trend reading will not be possible until next year. This is one of the most useful reasons to start the measurement work earlier rather than later. The three-quarter trend reading is only available to a leadership team that has been measuring consistently for three quarters.

Putting It Into Practice

This chapter named Phantom Productivity, traced where it lives, and showed how to tell it apart from legitimate setup friction. The work of this chapter is not complete until those concepts have been pointed at specific workflows inside the business. Before turning the page, run the following diagnostic on the three highest-activity AI workflows the business currently has.

For each of the three workflows, run through the four signals from earlier in the chapter. First, where has the AI-touched step gotten measurably faster, and can the answer be supported by timing data rather than impression? Second, has time spent reviewing, correcting, or revising AI outputs grown since AI was introduced? Third, are quality issues the upstream step appeared to eliminate now showing up further downstream, at the reviewer's desk or at the customer's? Fourth, has activity volume grown without a corresponding increase in completed, usable outputs? Three or four confirmations of Yes is high-probability Phantom Productivity. Two is probable. One or zero is likely real efficiency or legitimate setup friction.

Then move the diagnostic into action. Capture one representative week of upstream and downstream timing data for each workflow. The total time compared against the pre-AI baseline, or the reconstructed baseline if no documented one exists, is the read that confirms or denies the diagnosis. Categorize each workflow by signal count and let the category determine the next action rather than letting the team's intuition override it.

Open the three conversations described earlier in this chapter. The workflow-owner conversation first, focused on the operating evidence. The executive-sponsor conversation second, focused on the financial pattern and the decisions it implies. The P&L conversation with the CFO third, focused on where the cost is currently being absorbed and what the next quarter's measurement record will look like. None of the three is optional, and the order matters. Reversing the order surfaces conclusions before evidence and undermines the credibility of the discipline being installed.

Calendar the re-measurement before closing the diagnostic. Six weeks out is the right interval. Setup friction will have resolved by then; Phantom Productivity will not. The re-measurement, not the original reading, is the proof. A leadership team that runs the diagnostic once and never returns to it has produced a finding without producing a decision. A team that calendars the re-measurement before closing the first action has installed the operating discipline that the rest of this Volume builds on.

The next chapter puts a dollar figure on what Phantom Productivity is actually costing. Until that figure exists, every conversation about AI performance inside the business runs on impression. After the figure exists, the conversation runs on numbers. That shift is the operating beat the rest of the book builds on.

FIGURE **2.6**

Putting It Into Practice

Run the diagnostic on the three highest-activity AI workflows. The measurement, not the impression, is the finding.

THE FOUR PHANTOM SIGNALS — RUN AGAINST EACH HIGH-ACTIVITY WORKFLOW

S1 — Upstream Acceleration
Where has the AI-touched step gotten measurably faster? Time-tracking, prompt logs, or output timestamps should confirm it. A feeling rather than a number is itself a finding.

S2 — Downstream Review Burden
Has time spent reviewing, correcting, or revising AI outputs grown since AI was introduced? Track review minutes per output across a representative week.

S3 — Rework Reappearance
Are quality issues the upstream step appeared to eliminate showing up further downstream? An error type rare pre-AI now appearing at the reviewer's desk confirms it.

S4 — Capacity Disconnect
Has activity volume grown without a corresponding increase in completed, usable outputs? More drafts with flat completions is the capacity signal.

1 RUN THE UPSTREAM-DOWNSTREAM MEASUREMENT

Capture one representative week of timing data on each of the three workflows. Record upstream minutes and downstream minutes separately. The total compared against the pre-AI baseline (or the reconstructed baseline) is the diagnostic.

Owner: AI Efficiency Lead — within seven days

2 CATEGORIZE THE FINDINGS

For each workflow, label the result Phantom (three or four signals confirming Yes), Probable (two signals), or Real Efficiency (one signal or zero). The category determines the next action.

Owner: Workflow Owner with AI Efficiency Lead

3 OPEN THE THREE CONVERSATIONS

Schedule the workflow-owner conversation, the executive-sponsor conversation, and the P&L conversation with the CFO. None is optional. Order matters: workflow owner first, executive sponsor second, P&L third.

Owner: Executive Sponsor — within 14 days

4 CALENDAR THE RE-MEASUREMENT

Set the next measurement for six weeks out. Setup friction will resolve by then; Phantom Productivity will not. The re-measurement is the proof, not the original reading.

Owner: AI Efficiency Lead — schedule before closing the action

THE FOUR SIGNALS ARE THE DIAGNOSTIC. THE SIX-WEEK RE-MEASUREMENT IS THE PROOF.

Chapter 3: The AI Efficiency Tax™

The Hidden Cost That's Eating Your Margin

CHAPTER 3 OUTCOMES

At the end of this chapter, you will be able to:

1. Define the AI Efficiency Tax and name its five components: review burden, rework cycles, output inconsistency cost, management overhead, and shadow process cost.
2. Calculate a defensible loaded labor rate for each role used in AI-supported workflows, including the four SMB-specific complications.
3. Produce a Tax estimate per workflow and at the portfolio level, with documentation that holds up under CFO questioning.
4. Recognize why the Tax gets absorbed rather than surfaced inside most businesses and how to make it visible without blaming the team.
5. Read the quarterly Tax review and explain what changes between periods, why, and what corrective action follows.

THE AI EFFICIENCY TAX IS REAL DOLLARS. IT IS WHAT TURNS PHANTOM PRODUCTIVITY INTO A NUMBER YOU CAN DEFEND.

The Problem This Chapter Solves: You're Paying a Cost You're Not Tracking

Every AI-supported workflow in your business carries two price tags. The first one is easy to find. It's on the invoice. Subscription fees, seat licenses, platform costs, and any implementation work your team paid for. That number shows up on a budget line and gets reviewed every quarter. The second price tag is harder to find because it doesn't appear on any invoice. It lives inside your payroll, your manager schedules, your rework cycles, and the time your team spends cleaning up AI output before anyone can actually use it. That second price tag is the AI Efficiency Tax, and most businesses are paying it every single week without ever knowing it exists.

The AI Efficiency Tax is the total operational cost of making AI output usable.

It's the time a senior team member spends reviewing an AI-generated document before it goes to a client. It's the hour a manager loses tracking down inconsistent AI outputs from two different employees who used the same tool in two completely different ways. It's the rework cycle that opens up every time AI produces something that's structurally complete but factually wrong, contextually off, or missing the specific judgment that only a human can apply. It's the shadow process an employee built on the side because the approved AI workflow didn't actually work the way leadership thought it did. All of that cost is real. None of it shows up as a labeled expense.

Chapter 2 showed you what Phantom Productivity looks like from the outside. The work here shows you what it costs from the inside. Those two ideas are directly connected. Phantom Productivity is the pattern. The AI Efficiency Tax is the financial consequence of that pattern running unchecked through your operation. When AI shifts work rather than eliminating it, the shifted work lands somewhere. (George, 2024) It lands in review queues, correction cycles, manager inboxes, and informal workarounds that your best people are quietly running in the background. That work has a cost. And until leadership starts tracking it separately, it disappears into ordinary payroll and gets treated as a normal cost of doing business rather than a recoverable margin leak.

By the end of this chapter, you'll be able to estimate the hidden cost of your current AI workflows, identify which one is generating the largest margin leak, and understand exactly what to do about it before it becomes permanent.

The Five Components of the AI Efficiency Tax™

The AI Efficiency Tax isn't a single cost. It's five separate cost streams that move through your operation simultaneously, each one absorbed into a different part of your budget, and none of them labeled as AI-related. Understanding each component individually is the only way to measure the full tax, because if you look for it as one number, you'll miss most of it.

The first component is review burden. This is the time spent checking AI output before it can be trusted or used. Every AI-assisted workflow that requires a human to verify, approve, or inspect the output before it moves forward carries a review cost. The cost is real whether the review takes five minutes or forty-five. The problem isn't that review exists. The problem is that in most businesses, review time is never tracked as an AI-related cost. It gets absorbed into a manager's day, folded into a senior employee's schedule, or treated as a normal part of the job. When you add up the review hours across every AI-supported workflow in the business, the number is almost always larger than anyone expected. And because it's carried by senior people, it's priced at the highest labor rate in the room.

The second component is rework cycles. These are the repeated corrections caused by AI output that was incomplete, inaccurate, inconsistent, or missing critical context. Rework is different from review. Review is the check. Rework is what happens when the check fails and the output has to be corrected, regenerated, or rebuilt. Rework cycles are expensive because they consume time twice: once when AI produced the original output and once when a person had to fix it. In high-volume AI workflows, rework can accumulate fast. A team running fifty AI-generated outputs per week with a twenty percent correction rate is running ten rework cycles every week. At an hour each, that's ten hours of labor cost per week that the business is paying for AI to create and then paying again for a person to fix.

The third component is output inconsistency cost. This one is subtle but significant. When multiple employees use the same AI tool without shared standards, prompts, or quality criteria, the outputs vary widely. (Robert et al., 2020) One employee's AI-generated proposal looks completely different from another employee's. One team member's AI-assisted client report includes analysis that another team member's version omits entirely. That inconsistency creates downstream cost: client confusion, internal rework to standardize before delivery, management time spent reconciling different versions, and reputational risk when inconsistent quality reaches external stakeholders. Output inconsistency is a tax on the absence of standards, and it runs continuously until the business installs them.

The fourth component is management overhead. This is the time leaders and managers spend untangling questions, resolving inconsistencies, fielding employee confusion, and making judgment calls about AI outputs that the workflow never formally assigned to anyone. When a manager spends thirty minutes a day answering questions about whether an AI output is good enough to use, that's fifteen hours of management time per month. At a loaded labor rate of eighty dollars per hour, that's twelve hundred dollars per month in management cost that traces directly back to an AI workflow that wasn't designed clearly enough to run without constant supervision. Most businesses have multiple workflows generating this kind of overhead simultaneously, and none of it is tracked as AI cost.

The fifth component is shadow process cost, and it is the operating dollar figure that Volumes I, II, and III did not produce. Volume I Chapter 2 surfaced shadow AI as a category to bring into the light. Volume II Chapter 7 diagnosed it as an internal control issue scored on the AI Readiness Maturity Scale. Volume III Chapter 4 attached a breach cost to it through the IBM and Ponemon data. None of those Volumes calculated what an active, unbreached shadow process is costing the business in labor every month. That is the calculation this component produces. The shadow process cost is the waste created when employees build informal workarounds outside the approved AI process. Shadow processes appear when the official workflow doesn't work the way leadership thinks it does. (Semionovas, 2026) An employee discovers that the AI tool produces better results with a specific prompt sequence that wasn't in the training. Another employee skips the AI step entirely and uses a personal tool because it's faster. A third employee adds an undocumented review step because they've learned from experience that the AI output is unreliable in certain situations. Each of these workarounds represents time, inconsistency, and risk that the business is absorbing without knowing it. Shadow processes also mean that when leadership looks at workflow performance data, they're seeing numbers from a process that doesn't reflect what's actually happening on the ground.

Each of these five components has a cost you can estimate. Time spent, multiplied by how often it happens, multiplied by the loaded labor rate of the person doing it. That formula works for all five. The loaded labor rate should include salary, benefits, and overhead, which typically runs between 1.25 and 1.4 times the base salary for most SMB roles. (Ooi & Goh, 2018) Use whichever rate your

CFO would use. The goal isn't a perfect number. The goal is a directionally honest number that changes the leadership conversation from "AI is probably helping" to "AI is costing or saving us this much, and here's the math."

Calculating the Loaded Labor Rate in an SMB Context

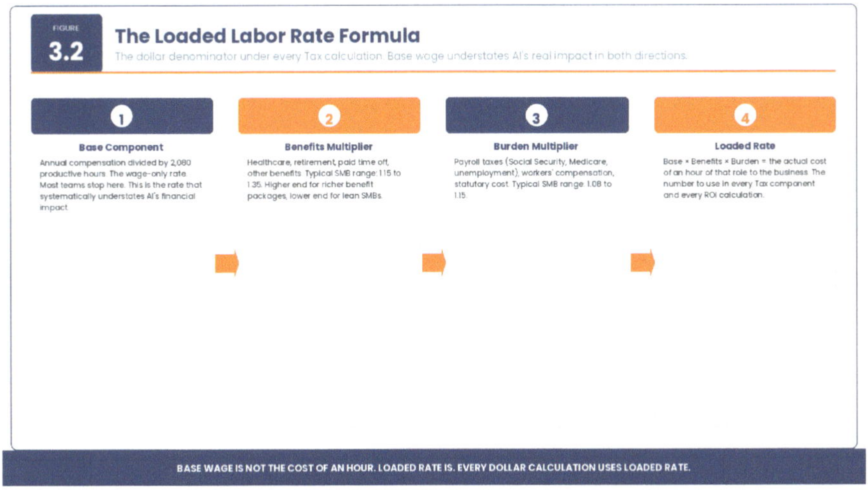

The tax components above all reduce to the same multiplication: time spent, times frequency, times loaded labor rate. The first two are observable. The third requires a specific calculation that often gets done badly in SMB contexts, and getting it wrong distorts the entire tax estimate. This section gives you a working methodology for producing a defensible loaded labor rate without rebuilding your payroll model.

The standard formula for an employee on W-2 payroll is straightforward. Base salary, divided by working hours per year, multiplied by a loading factor between 1.25 and 1.4 to capture benefits, payroll taxes, workspace cost, and overhead. For most professional roles in most SMBs, working hours per year run between 1,800 and 2,000 once vacation and holidays are removed. (Hanes, 2012) A team member on a $90,000 base salary at 1,900 working hours produces an unloaded rate of approximately $47 per hour. At a 1.3 loading factor, the loaded rate is about $61. Use the loading factor your CFO already applies to internal cost models. If your business does not have a documented factor, 1.3 is a defensible default for most professional service contexts and 1.25 is a defensible default for production or operational roles.

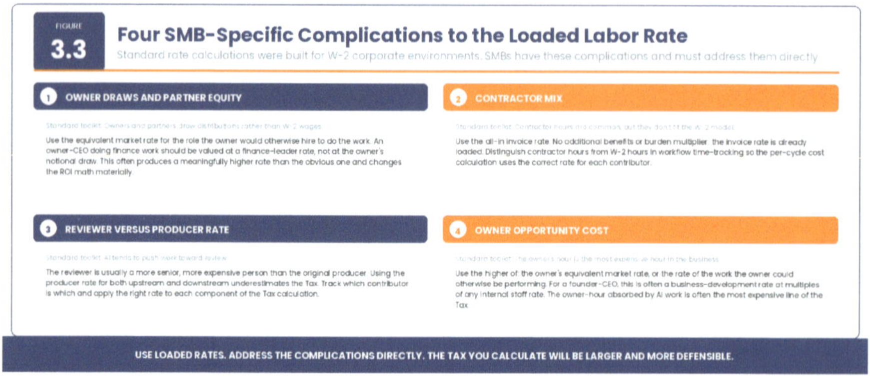

The SMB-specific complications are worth handling explicitly. The first complication is owner draws and partner equity. When the partner reviewing an AI output is paid through draws rather than salary, the implied hourly rate is what the partner could otherwise be billing or what a comparable hire on payroll would cost. For partners in professional services firms, that number is usually significantly higher than the staff loaded rate, often $150 to $300 per hour depending on the practice. (Maister, 1982) Use the comparable-hire rate, because that is the rate the business would have to pay to replace the partner's review time with someone on payroll, and that is the financially honest measure of what the review is costing.

The second complication is contractor or part-time mix. Contractors are typically priced at their full billable rate without a loading factor, because the rate already includes the contractor's own overhead. Part-time employees use the same loading factor as full-time but with their actual scheduled hours rather than a 1,900-hour assumption. A part-time analyst at $40,000 working 1,000 scheduled hours has an unloaded rate of $40 per hour and a loaded rate of approximately $52 at a 1.3 factor.

The third complication is the labor cost of the reviewer versus the producer. AI Efficiency Tax™ components like review burden and management overhead are carried by senior people whose loaded rates are significantly higher than the staff who run the AI tool. Calculate each component at the rate of the person actually doing that component's work, not the rate of the workflow's average employee. A workflow where an analyst at $50 per hour produces output that a partner at $200 per hour reviews has a tax rate weighted heavily toward the partner's time, and reporting it at the analyst's rate understates the tax by a factor of four.

The fourth complication is the cost of the owner's own time. In small businesses where the CEO, COO, or managing partner personally reviews AI output or resolves AI-related questions, that time has a real opportunity cost that does not appear on a payroll line. Estimate it at the loaded rate of the senior hire who would otherwise be doing that work, or at a defensible benchmark for the role. Owner time is real labor cost even when it is not on payroll, and leaving it out of the tax calculation systematically understates what AI is actually costing the business.

A working approach for an SMB leadership team that does not have time to build a formal labor cost model is to produce a short table of loaded rates by role: staff, senior staff, manager, director, partner or owner. Three to five rows is enough for most SMB structures. Use that table consistently across every tax calculation in the business. The goal is not perfect precision. It is a defensible, consistent rate methodology that produces tax estimates the CFO will recognize as financially honest and that hold up when someone in the room asks how the number was built.

Why the AI Efficiency Tax Gets Absorbed and Never Surfaced

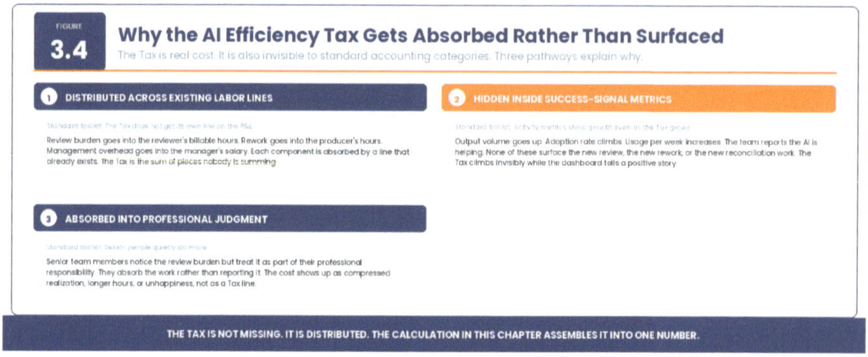

The AI Efficiency Tax stays invisible because the business pays it through operating capacity rather than through a separate expense category. There's no invoice line that says "AI review burden" or "AI rework cost." The cost enters through payroll, manager time, and workflow friction, and it gets treated as ordinary operating expense. That's the structural reason the tax stays hidden. But there are three specific patterns that keep it invisible even when leadership is actively trying to understand AI performance.

The first pattern is the absence of a pre-AI baseline. When AI is introduced into a workflow without capturing what that workflow cost before, the business has no reference point for measuring what changed. Review time that existed before AI was introduced looks the same as review time that AI created. (Oke, 2008) Rework that AI generated looks identical to rework that was always there. Without a baseline, there's no way to separate the AI-related cost from the pre-existing cost, so the entire cost gets treated as normal and nobody connects it back to the tools. This is one of the most common and most expensive measurement failures in AI adoption, and it's almost entirely preventable. Capturing a baseline before introducing AI into a workflow takes less than a day. Not capturing one costs the business months of invisible expense.

The second pattern is the absence of an AI cost label. When review time, rework, and correction cycles aren't tracked as AI-related costs, they get absorbed into general labor and overhead. A manager who spends two hours a week reviewing AI outputs doesn't log that time as "AI review." She logs it as "client work" or "team support" or nothing at all. A team member who spends forty minutes correcting an AI-generated report doesn't categorize that time as "AI rework." He treats it as part of his normal workday. Because the cost is never labeled, it never aggregates. It never becomes visible as a category. And because it's never visible as a category, it never gets addressed as a problem. The tax runs continuously, absorbed into the payroll line, completely invisible to anyone reviewing the budget.

The third pattern is the absence of an owner. In most businesses, nobody is specifically responsible for tracking the cost of making AI output usable. The CFO owns the budget. The COO owns operations. The department heads own their workflows. But the specific question of how much time and money the business is spending on review, correction, and management of AI outputs falls between those responsibilities. Nobody is assigned to track it. Nobody is accountable for reducing it. And because no one owns it, no one measures it, which means no one reports it, which means leadership keeps believing that AI is saving time while the organization feels just as busy as it did before the tools were introduced.

This is the mechanism behind the disconnect that many leadership teams experience. The CEO sees high AI usage and assumes efficiency is improving. The COO hears from department heads that the tools are helpful. But the team feels overloaded. Managers are stretched. Capacity hasn't opened up. The work isn't getting done faster. That gap between what leadership sees in the usage data and what the organization feels in its daily operations is the AI Efficiency Tax running undetected. The work wasn't eliminated. It was redistributed into places the business doesn't measure, carried by people who don't have a category to log it under, and absorbed into a budget that has no line for it.

The practical correction is straightforward, and it doesn't require a new system or a data team. For every major AI-supported workflow, require thirty days of separate tracking for three specific numbers: time spent on review, time spent on correction, and time spent on exception handling when the AI output can't be used as delivered. Track those three numbers by workflow, by person, and by frequency. At the end of thirty days, total them up. Multiply by loaded labor rate. That number is the AI Efficiency Tax for that workflow. It won't be a perfect number. It will be an honest one. And an honest number, even a rough one, changes the conversation in a way that no usage report ever does.

Calculating Your Current AI Efficiency Tax™

The calculation doesn't require a data science team, a business intelligence platform, or a complex model. It requires three inputs per workflow: time, frequency, and loaded labor rate. Everything else is multiplication and addition. The goal is a directionally honest estimate, not an audit-grade financial statement. A number that's eighty percent accurate and available this week is worth more than a number that's perfect and available next quarter. (Brownstone, 1996)

Start with your top three active AI use cases. These are the workflows where AI is most embedded, most frequently used, and most likely carrying a significant tax. For each one, work through the five tax components in order.

For review burden, ask: who reviews AI output in this workflow before it can be used, how long does that review typically take, and how many times per week does that review happen? Multiply time by frequency to get weekly hours. Multiply weekly hours by loaded labor rate to get weekly cost. Multiply by 52 for an annual figure.

For rework cycles, ask: how often does AI output require correction, regeneration, or significant editing before it's usable? Estimate the average time per correction cycle and the number of correction cycles per week. Same multiplication: time times frequency times rate equals weekly cost.

For output inconsistency, this one is harder to quantify directly, but a reasonable estimate comes from asking how much time per week is spent reconciling, standardizing, or aligning AI outputs across team members before delivery. Include the management time spent addressing quality concerns that trace back to inconsistent AI outputs.

For management overhead, ask each manager who touches an AI-supported workflow how much time per week they spend answering AI-related questions, resolving AI output issues, or making judgment calls about whether an AI output is acceptable. That time, multiplied by their loaded labor rate, is the management overhead component of the tax for that workflow.

For shadow process cost, this requires a direct conversation with the team. Ask specifically: is there any part of this workflow where people are doing something different from the official process? Where are they adding steps, skipping steps, or using different tools than the ones we approved? Estimate the time those shadow steps consume per week and add them to the total.

To make this concrete, consider a hypothetical scenario. A professional services firm is using AI to generate first drafts of client-facing reports. The AI saves each analyst about ninety minutes of drafting time per report. That looks like a clear win. But when a senior partner reviews each report before delivery, that review takes about an hour. When the analyst corrects the issues the partner identifies, that takes another thirty minutes. The AI tool produces inconsistent quality depending on which analyst runs it, so the partner is spending an extra fifteen minutes per report flagging style and format issues that wouldn't exist if there were shared prompting standards. Total time saved per report: ninety minutes. Total tax per report: one hour and forty-five minutes of review and correction, carried by a senior partner at a significantly higher labor rate than the analyst who ran the AI tool. The firm is spending more on the tax than it saved on the draft. The AI use case is running at a net loss, and it looks like a productivity win on every usage report the firm produces.

That's the kind of clarity this calculation creates. The number doesn't need to be perfect. It needs to be honest enough to change what leadership decides to do next. Once you have a monthly and annual estimate for each of your top three workflows, rank them by total tax cost. The workflow with the highest tax is the one that deserves immediate attention, not because it's the worst-run workflow in the business, but because it's the one where reducing the tax will recover the most margin in the shortest amount of time.

From Tax to Savings: How Reducing the AI Efficiency Tax™ Directly Improves Margin

Reducing the AI Efficiency Tax is margin recovery. That framing matters because it changes how leadership treats this work. It's not a process improvement project. It's not an IT initiative. It's a financial recovery effort with a specific dollar amount attached to it, and it produces results that show up in the P&L without requiring new headcount, new tools, or new capital investment.

Every hour of review burden that gets eliminated gives a senior employee or manager that hour back. Every rework cycle that gets removed gives the business the labor cost of that cycle back. Every shadow process that gets replaced by a clean, standardized workflow eliminates the time cost of running the workaround and reduces the inconsistency cost it was generating downstream. These aren't theoretical savings. They're real labor hours that were being consumed by a tax that nobody was tracking, now freed up for higher-value work because the tax has been reduced or removed.

The sequence that produces the fastest margin recovery is specific. Before leadership approves any new AI use cases or expands AI into additional workflows, it should first recover the margin being lost in the workflows already using AI. This is the most important strategic insight in this chapter, and it's the one most often skipped. Expanding AI into new workflows while existing workflows are running a significant tax is the equivalent of opening a second location while the first one is losing money. The new activity creates the appearance of growth while the underlying cost problem compounds.

The first efficiency win is almost always found in the current operation, not in a new use case. The cheapest AI gain available to most businesses right now is fixing the waste that already exists inside the workflows AI is already touching. That work doesn't require a new budget line, a new vendor, or a new implementation project. It requires a clear-eyed look at where the tax is highest, a decision about what's causing it, and a structured workflow correction that reduces or removes the cost.

The three most common tax reduction moves are standardization, checkpoint redesign, and use case removal. Standardization means creating shared prompting guidelines, output quality criteria, and review standards so that every team member using the same AI tool produces consistent results that require less correction and less management oversight. Checkpoint redesign means moving the quality check to an earlier point in the workflow so that problems are caught before they compound into expensive rework at the end. Use case removal means acknowledging that some AI applications are generating more tax than value and removing them from the workflow entirely, which immediately recovers the labor cost of managing them.

Each of these moves produces a measurable financial result. When review burden drops from sixty minutes per output to fifteen minutes, that's forty-five minutes of senior labor recovered per output. Multiply that by weekly output volume and you have a weekly savings number. Multiply by 52 and you have an annualized figure that belongs in the AI ROI Formula covered later in this book. That number is defensible. It's based on measured time, documented frequency, and a loaded labor rate that any CFO already uses. It's not an estimate of future potential. It's a measurement of past waste that the business has now recovered.

The CFO case for this work is straightforward. Tax reduction converts hidden labor leakage into documented savings without adding cost. The investment is leadership attention and a structured workflow review, not capital. The return is recovered margin, freed capacity, and a defensible efficiency number that can be reported to any stakeholder who asks what AI has actually done for the business. That's a strong financial argument, and it gets stronger every quarter the business runs a lower tax rate on its AI workflows.

The AI Efficiency Tax in Multi-Use-Case Operations: Avoiding Double Counting

Most SMBs run multiple AI use cases that share underlying infrastructure. The same subscription supports drafting, analysis, and customer communication. The same staff reviewers handle output from several workflows. The same management overhead covers governance for the entire AI program rather than for any single use case. When the AI Efficiency Tax is calculated separately for each use case without addressing this overlap, the same cost gets counted multiple times, and the aggregate Tax figure presented to leadership is overstated by a factor that the CFO will identify in minutes. Correct multi-use-case Tax calculation requires explicit allocation methodology.

Three categories of cost typically need allocation rather than full assignment. Subscription cost is the most obvious. A single AI platform subscription supporting five use cases should not be charged at full cost against each of the five. The defensible methods are allocation by usage volume (which use case generated the most AI activity in the period), allocation by labor savings (which use case produced the most measured benefit), or proportional allocation (each use case bears one-fifth of the subscription cost). The right method depends on what the calculation is being used for. For aggregate Tax reporting, proportional allocation is usually cleanest. For per-use-case ROI calculation, usage-based or savings-based allocation is more accurate. Document the method and apply it consistently across the calculation.

Shared review labor is the second category. A senior reviewer who spends ten hours per week reviewing AI output across three workflows should not have that ten hours counted three times in the Tax calculation. The defensible method is to ask the reviewer to estimate how their review time splits across the workflows in a typical week, document the percentages, and allocate the loaded labor cost accordingly. A reviewer who reports forty percent on workflow A, forty percent on workflow B, and twenty percent on workflow C produces clean allocation that the Tax calculation can use without double-counting.

Management overhead is the third category. The time leadership spends on AI program governance, vendor management, and incident response is typically allocated across all active AI use cases rather than charged against any single one. A small fraction of management overhead is sometimes specific to a particular use case (when a single workflow has produced an incident, for example), but the bulk of it covers the program. Allocate proportionally across active use cases unless there is a specific reason to weight the allocation differently.

The output of this methodology is two views of the Tax figure for any given use case. The unit Tax is what the use case is costing in isolation, calculated on its own dedicated labor and proportional share of shared costs. The total Tax is the sum of unit Taxes across the AI program. Both views are useful. The unit Tax supports per-use-case decisions about whether to expand, refine, pause, or remove. The total Tax supports program-level decisions about whether the AI investment as a whole is producing acceptable net value. A leadership team that presents both views to its CFO will have a more credible conversation than one that presents either view alone.

Quarterly Tax Review: What Changes Between Periods and Why

The AI Efficiency Tax is not static. It moves quarter to quarter for specific operational reasons, and the leadership team that reviews the Tax each quarter develops a sharper feel for which movements signal real change and which are noise. The most common quarterly movements come from three sources: seasonal volume changes, team turnover, and tool changes.

Seasonal volume changes move the Tax in the direction of the volume change. If the workflow produces more units in the high season, the Tax components that scale with volume (review burden, rework cycles, shadow process cost) move up. If the team is unchanged through the high season, the per-unit Tax may actually decline because fixed cost components (management overhead) are spread over more units. Reading the Tax against volume rather than in absolute dollars produces the clearer picture. A Tax that has grown ten percent in a quarter when volume has grown thirty percent is actually a per-unit Tax that has declined, which is a positive operational signal even though the absolute number is higher.

Team turnover moves the Tax in patterns that depend on what role turned over. When a senior reviewer leaves, the Tax usually spikes for the quarter in which their replacement is learning the work, then declines as the replacement reaches productivity. When a producer leaves, the Tax often declines briefly because the workflow's output drops, then returns to normal once the replacement is producing. When the AI program owner leaves, the Tax may not move in the immediate quarter but tends to drift upward in subsequent quarters because the program loses the active management that was keeping the Tax from compounding. Reading turnover effects on the Tax is one of the more useful inputs to staffing decisions in AI-heavy functions.

Tool changes are the largest source of unexpected Tax movement. Vendor product updates can change the AI tool's output quality, which moves the review burden and rework cost components. A new feature can reduce the Tax for one workflow while introducing it for another that the new feature is now reaching. A pricing change moves the subscription cost component. A model upgrade often moves all five components in different directions simultaneously, and the net effect on the Tax can take a full quarter to settle. The discipline is to capture the Tax components separately rather than only as an aggregate, so that a tool change that moves three components in different directions can be read accurately rather than only seen as a single number that moved a little.

The quarterly Tax review meeting is the operational instrument that converts these movements from noise into signal. The leadership team reviews the Tax across all active AI use cases, separates the components, identifies the movements quarter over quarter, and tags each movement with a probable cause. The output is a documented quarterly Tax record that, after four quarters, becomes one of the most useful operating diagnostics in the AI program. A leadership team that has run the quarterly Tax review for a year can read trends that the same team would have missed without the cadence.

What a Defensible AI Efficiency Tax Calculation Looks Like to a CFO

The Tax calculation is going to land in front of a CFO at some point. Either the leadership team brings it to the CFO during budget review, or the CFO discovers it in the AI ROI Formula calculation during financial planning, or the Tax shows up in the Executive AI Efficiency Brief when the Brief is presented to the board. The CFO will examine the calculation against the same standards the finance team applies to any other operating cost analysis, and a Tax calculation that fails those standards loses credibility for the entire AI program. Three components determine whether the calculation will hold up under CFO scrutiny.

The first is the loaded labor rate. CFOs know what loaded labor rates look like in the business, and a Tax calculation that uses an unloaded hourly rate, an outdated rate, or a rate that does not match the role mix actually doing the work will be flagged immediately. The defensible posture is to use the same loaded labor rate methodology the CFO already applies to other internal cost models, sourced from the same payroll data, with the documented loading factor the business uses elsewhere. If the CFO cannot find a difference between the Tax calculation rate methodology and the rate methodology used in standard operating cost analyses, the rate component is defensible.

The second is the cost denominator completeness. A Tax calculation that includes only subscription cost while ignoring review burden, rework cycles, output inconsistency, management overhead, and shadow process cost will be identified as incomplete in the first review. The CFO will reasonably ask about each missing component. The defensible posture is to present all five components even when some are estimated rather than directly measured, with the methodology for each estimate documented and the range of plausible values noted. A complete calculation with documented estimates beats a partial calculation with precise numbers every time, because the complete calculation invites discussion while the partial calculation invites distrust.

The third is the connection to specific workflows. A Tax calculation reported at the program level without traceability to individual use cases will be questioned. A CFO who is asked to act on a Tax figure will ask which workflows are generating the largest components, what the leadership team is doing about them, and what the Tax would look like if the corrective actions were successful. The defensible posture is to present the Tax with per-use-case breakdowns, identified largest contributors, named corrective actions in flight, and expected forward Tax trajectory if those actions succeed. The CFO is then equipped to act on the calculation rather than only to critique it.

A Tax calculation that meets all three standards becomes one of the most credible operating reports the AI program produces. CFOs who have seen a defensible Tax calculation usually trust the rest of the AI Operating System artifacts more readily, because the discipline behind the Tax calculation signals the discipline behind the program. CFOs who have seen an indefensible Tax calculation often become the first source of resistance to the AI program more broadly, because the partial calculation signals partial discipline elsewhere. The investment in producing a defensible Tax is one of the highest-impact uses of the leadership team's time in the first six months of the Volume IV work.

The operating principle that ties this together is simple. The first AI efficiency win isn't a new use case. It's making the current use case stop quietly consuming the value it was supposed to create. Fix the tax before expanding the investment. Recover the margin before adding the next tool. That discipline is what separates a business that uses AI from a business that benefits from AI, and the difference between those two things shows up directly in the P&L.

Putting It Into Practice

The AI Efficiency Tax becomes manageable the moment it becomes visible. Right now, for most businesses, it's invisible because it's absorbed into ordinary labor, manager schedules, and workflow friction with no label attached. The work in this chapter is about attaching the label, estimating the cost, and identifying where the tax is highest so the business can address it in order of financial impact.

Here are four specific steps to complete before moving into the next chapter.

First, select your top three active AI use cases by workflow activity. These are the workflows where AI is most embedded and most frequently used. Write them down by name, not by tool. "AI-assisted client reporting workflow" is more useful than "ChatGPT" because the tax lives in the workflow, not the software.

Second, for each workflow, estimate the five tax components using the time, frequency, and loaded labor rate method described in this chapter. You don't need perfect data. You need honest estimates from the people closest to the work. Schedule a thirty-minute conversation with one team member per workflow and ask them directly: where are you spending time reviewing, correcting, or working around AI outputs in this process? Their answers will get you to a working estimate faster than any internal report will.

Third, calculate the monthly and annual cost for each workflow by totaling the five components, each one already calculated as time multiplied by frequency multiplied by loaded labor rate, and projecting forward. Write the number down.

Having a specific dollar figure, even a rough one, changes the conversation in every leadership meeting where AI performance comes up. A CFO who hears "our AI reporting workflow is generating approximately four thousand dollars per month in review and rework cost" responds differently than a CFO who hears "we think there might be some inefficiency in the reporting process."

Fourth, rank your three workflows by total tax cost and identify which one is generating the largest margin leak. That workflow is your first optimization target. Don't try to fix all three at once. The business has limited leadership attention and limited change capacity. Prioritize the workflow where the tax is highest, the business impact is greatest, and the fix is most practical to implement. A focused correction on one high-tax workflow will recover more margin faster than a diffuse effort spread across all three.

Once you have that ranking, you know exactly where the money is going and which workflow to address first. The next chapter introduces the Workflow Reality Map, which gives you the tool to look inside that high-tax workflow, find exactly where each component of the tax is entering the process, and design a correction that removes the cost at the source rather than managing it after the fact. The tax estimate you build here becomes the input that makes the Workflow Reality Map precise, targeted, and financially meaningful from the first day you use it.

FIGURE 3.6

Putting It Into Practice

Attach a dollar figure to the Tax in four steps. Honest estimates from the people closest to the work beat any internal report.

THE FOUR INPUTS YOU NEED BEFORE THE FIRST CALCULATION

I1 — Workflow name, not tool name
The Tax lives in the workflow, not the software. "AI-assisted client reporting workflow" is more useful than "ChatGPT." Name the work, not the vendor.

I2 — Time per occurrence
Minutes per review, per rework cycle, per reconciliation, per shadow-process instance. From direct observation or the thirty-minute conversation with one team member per workflow.

I3 — Frequency
Occurrences per week for each component. Multiplied by 52 for the annualized number. Frequency is where most underestimation hides; capture it honestly.

I4 — Loaded labor rate
Base wage × benefits multiplier × burden multiplier. Producer rate for some lines, reviewer rate for others. Owner draws priced at the equivalent market rate.

1 SELECT THE TOP THREE USE CASES

Pick the three AI-supported workflows where AI is most embedded and most frequently used. Write them down by workflow name, not by tool. The Tax lives in the workflow, not the software.

Owner: AI Efficiency Lead — within seven days

2 ESTIMATE THE FIVE TAX COMPONENTS

For each workflow, run the time × frequency × loaded labor rate method on review burden, rework cycles, output inconsistency cost, management overhead, and shadow process cost. Honest estimates from people closest to the work beat any internal report.

Owner: Workflow Owner with AI Efficiency Lead

3 CALCULATE MONTHLY AND ANNUAL COST

Total the five components for each workflow. Project forward to monthly and annual. Write the number down. A specific dollar figure, even a rough one, changes every leadership conversation where AI performance comes up.

Owner: AI Efficiency Lead with CFO support

4 RANK AND PICK THE FIRST TARGET

Order the three workflows by total Tax. The highest one is the first optimization target. Don't try to fix all three at once. A focused correction on the largest margin leak recovers more margin faster than a diffuse effort across all three.

Owner: Executive Sponsor — decision at next operating meeting

THE NUMBER YOU WRITE DOWN HERE BECOMES THE INPUT THAT MAKES THE WORKFLOW REALITY MAP IN CHAPTER 4 PRECISE.

Chapter 4: The Workflow Reality Map

Seeing Your AI Workflows as They Actually Are

CHAPTER 4 OUTCOMES

At the end of this chapter, you will be able to:

1. Distinguish the workflow as designed from the workflow as actually run, and explain why the gap between the two is where AI lives.
2. Build a Workflow Reality Map for a priority AI-supported workflow using the four-step protocol.
3. Mark each step with one of four findings: Gain, Loss, Risk, or Unknown — and defend the label with evidence.
4. Identify the AI touchpoint type at each step: Producer, Reviewer, Augmenter, or Replacer.
5. Recognize the six recurring workflow patterns that show up across SMB AI operations and the response each one calls for.

THE MAP IS NOT DOCUMENTATION. IT IS A DIAGNOSTIC INSTRUMENT THAT TURNS A WORKFLOW INTO EVIDENCE.

Most businesses that have reached this point in the book have already done something valuable. They've named the AI Efficiency Gap. They've identified where Phantom Productivity is hiding. They've estimated the AI Efficiency Tax on their highest-activity workflows. Those three steps create clarity that most organizations never reach. But clarity about what a problem costs is different from clarity about where the problem lives. To fix the tax, you need to find it inside the actual workflow, step by step, handoff by handoff, at the level where the real work moves through real people in real time.

That's what this chapter is built to do.

The tool is called the Workflow Reality Map. The name is intentional. Not a workflow diagram. Not a process flowchart. A reality map, because the most important word in the name isn't "workflow" or "map." It's "reality." The map only works if it shows the workflow as it actually runs, not as it was designed, not as leadership believes it runs, and not as the official documentation says it should run. The gap between those two versions of the same workflow is exactly where AI gets placed incorrectly (Kumar, 2023) , where efficiency disappears, and where the AI Efficiency Tax accumulates every single week.

The Problem This Chapter Solves: You're Optimizing a Workflow That Doesn't Exist

When a business documents its workflows, it almost always documents the intended version. The process that was designed in a planning meeting, approved by a manager, and written into a procedure document at some point in the past. That documented version shows clean handoffs, clear approvals, logical sequencing, and a tidy path from start to finish. It looks organized because it was organized, at the moment it was created, by people who were describing how they hoped the work would move.

But that moment was in the past. And workflows drift.

They drift because the business changes and the documentation doesn't. (Ellis et al., 1995) They drift because an employee found a faster way to do something and started doing it that way without telling anyone. They drift because a software tool changed, a vendor changed, or a client requirement changed, and the team adapted on the fly without updating the official process. They drift because an exception happened once, then happened again, and eventually became so common that the team built an informal workaround that now runs parallel to the approved process without anyone in leadership knowing it exists.

By the time AI enters the picture, the real workflow may already include informal approval steps that don't appear anywhere in the documentation, manual data transfers between systems that were supposed to talk to each other automatically, spreadsheet patches built by someone who needed a fix and never got a better solution, shadow tools that individual employees prefer over the approved platform, and exception handling routines that consume significant time every week but have never been formally acknowledged as part of the process. None of that shows up in the documented version. All of it affects how AI performs inside the workflow.

When AI gets placed into the documented version of a workflow rather than the real one, the consequences are predictable. The tool may accelerate a step that wasn't actually the bottleneck. It may produce output that enters the process at the wrong point, creating handoff friction that nobody anticipated. It may skip over an informal approval step that employees were doing manually, causing rework when the output reaches someone who expected that step to have happened. Or it may interact cleanly with the official process while completely missing the shadow process running alongside it, leaving the real source of delay, cost, and inconsistency untouched.

This is why AI optimization so often produces disappointing results even when the tool itself is excellent. The tool is being optimized for a workflow that doesn't exist in practice. The real workflow, with all its informal patches and undocumented detours, keeps running underneath the optimized version, carrying the same inefficiencies it always carried, and generating the same costs it always generated. Leadership looks at the AI performance data and sees improvement in the step they optimized. The team feels no meaningful change because the actual friction points were never touched.

The Workflow Reality Map fixes this by making the real workflow visible before any optimization decision is made. The goal is simple and non-negotiable: before you optimize a workflow with AI, you must first see the workflow that actually exists.

Designed Workflows vs. Real Workflows. Why the Gap Matters

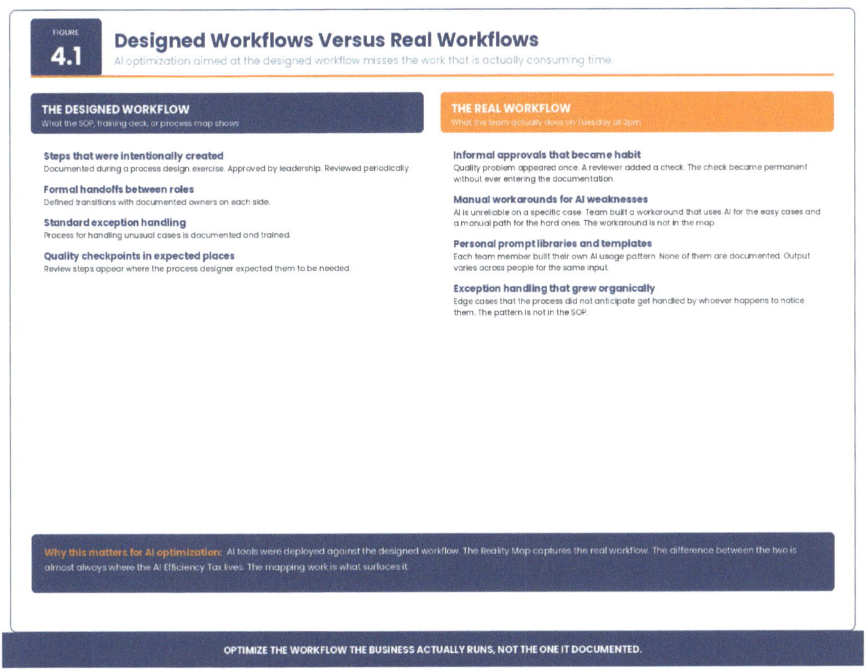

Every workflow in your business exists in two versions simultaneously. The designed version and the real version. Understanding the difference between them isn't an academic exercise. It's the operating foundation that determines whether AI creates efficiency or adds cost.

The designed version is the official process. It's what appears in the procedure manual, the onboarding documentation, the process diagram on the shared drive, or the workflow your operations team built in whatever project management software the company uses. It shows who does what, in what order, with what approval at each stage. It's clean, logical, and usually built by people who understood the work at the time they designed it.

The real version is what the team actually does every day. It includes everything the designed version doesn't show: the informal check a senior employee does before a deliverable moves forward because they've learned from experience that the next stage will reject it if certain things aren't confirmed first. The manual copy-paste between two systems that were supposed to integrate but never did. The approval that technically requires a manager's sign-off but in practice gets a text message confirmation because the manager is rarely available during the window when the work needs to move. The exception handling that one specific employee does because she's the only person who knows how to deal with a particular client's requirements, and nobody has ever written that knowledge down.

Workflows drift from their designed versions through a predictable set of forces. The most (Kammer et al., 2000) common is informal approval creep. A quality problem happens once. A manager adds a review step to prevent it from happening again. The review step becomes habit. It never gets added to the official documentation, but it runs in every cycle from that point forward, consuming time and creating a handoff that nobody has formally assigned to anyone. The second force is manual patching. A system fails to do something automatically, so an employee builds a workaround. The workaround works well enough that nobody fixes the underlying system problem. The workaround becomes permanent, invisible in the documentation, consuming labor hours in every cycle.

The third force is shadow tool adoption. An employee finds a tool that works better for a specific task than the approved platform (Mallmann et al., 2018) . They start using it quietly. Others notice and start using it too. Within a few months, a significant portion of the team is running a workflow step through an unapproved tool that leadership doesn't know about, doesn't govern, and isn't measuring. The fourth force is exception normalization. An exception happens often enough that the team builds a routine around handling it (Strong & Miller, 1995) . The routine becomes an unofficial sub-process. It consumes real time in every cycle where that exception appears, but it's never acknowledged in the official workflow documentation, so leadership has no visibility into how often it happens or what it costs.

When AI is added to a workflow without understanding these forces, the tool operates in a version of the workflow that doesn't match the version the team is actually running. The AI can't improve the informal approval step it doesn't know exists. It can't eliminate the manual data transfer that nobody documented. It can't reduce the exception handling cost that never appears in the official process. And if AI is placed at a step that looks like a bottleneck in the designed version but isn't actually the bottleneck in the real version, the tool creates faster output at a step that wasn't slowing anything down while the true source of delay continues running untouched.

AI cannot improve the workflow leadership imagines. It can only affect the workflow employees actually use. (Quaquebeke & Gerpott, 2023) That distinction is the entire reason the Workflow Reality Map exists, and it's the reason this chapter comes before any discussion of AI placement decisions or optimization choices.

How to Build a Workflow Reality Map in Four Steps

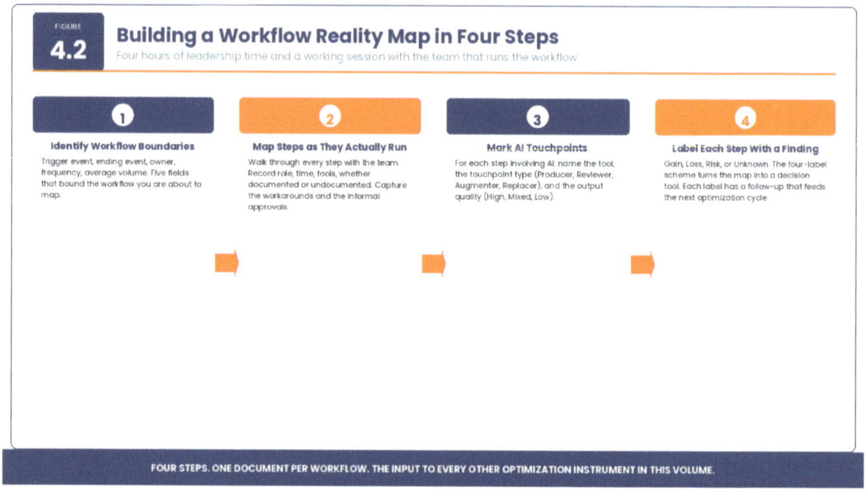

The Workflow Reality Map is a practical tool a lean leadership team can build without consultants, without specialized software, and without a complex analytics infrastructure. It requires time, honest conversation, and a willingness to see what's actually happening rather than what the documentation says should be happening. The four steps below are specific and sequential. Each one builds on the previous, and skipping any of them produces a map that misses the exact information it was designed to capture.

Start with the workflow that carries your highest AI Efficiency Tax from Chapter 3. That's the right place to begin because the tax estimate already tells you that something is generating hidden cost in that workflow. The Reality Map will show you exactly where that cost is entering the process and why.

Step one is to observe the workflow as it actually runs. This means sitting down with the people who do the work, not the people who designed the process or manage it from above, but the people who execute it every single day. Ask them to walk you through what they actually do from the moment a task enters the workflow to the moment it exits as a completed output. Ask open questions. What happens first? Then what? Who does that? What do you do when that doesn't work the way it's supposed to? What happens when the AI output isn't usable? Who checks it? What do you do with it after that? Listen for the words "usually" and "sometimes" and "it depends." Those words signal the informal steps, the exceptions, and the workarounds that don't appear in any documentation. Write down everything. Don't filter it for relevance during this step. Filtering comes later. Right now, the goal is a complete, unedited picture of what the team actually does.

If possible, observe the workflow in real time rather than relying entirely on description. Watch one complete cycle from start to finish. Note every tool that gets opened, every system that gets accessed, every person who touches the work at any stage. Note the moments where the person doing the work pauses, switches to a different tool, sends a message to someone else, or does something that wasn't part of the description they gave you. Those moments are where the real workflow diverges from the described version, and they're often where the highest cost is hiding.

Step two is to document every step, handoff, approval, delay, exception, manual transfer, and workaround, even the ones that feel too small or too informal to matter. This is the step where most workflow documentation efforts fail, because the person doing the documenting decides in real time which steps are "real" steps worth recording and which ones are minor details not worth capturing. That filtering instinct is wrong for this purpose. The informal approval that takes two minutes matters. The manual copy-paste that happens three times per cycle matters. The text message confirmation that substitutes for a formal sign-off matters. The exception that happens in one out of every five cycles matters. These are not side details. They're often the exact points where AI creates friction, where review burden accumulates, and where the AI Efficiency Tax enters the process.

Document the workflow in a simple visual format. A linear sequence of boxes connected by arrows works well for most workflows. Each box represents one step. Each arrow represents a handoff or a transition. Add a separate color or symbol for steps that are informal, undocumented, or inconsistently performed. Add another color or symbol for exception handling steps. Add a third for manual workarounds. When you're done, the map should show the full real workflow, including all the unofficial steps and informal patches, in a format that anyone on the leadership team can read in five minutes.

Step three is to mark every point where AI touches the workflow. For each AI touchpoint, capture six specific pieces of information: who uses the AI tool at that step, which tool or platform they use, what input they give the tool, what output the tool produces, who reviews that output, and what happens to the output after the review. These six data points are what transform a generic workflow map into an AI-specific diagnostic tool. They show exactly where AI is operating, what it's producing, and how the organization handles what it produces. Without these six data points, the map can show you where a workflow is slow or expensive, but it can't show you whether AI is causing the problem, contributing to it, or operating cleanly while the problem lives somewhere else entirely.

Pay particular attention to the gap between what the AI tool produces and what the next step in the workflow requires. If AI produces a draft document and the next step requires a fact-checked, client-specific deliverable, the gap between those two things is where review burden lives. If AI produces a data summary and the next step requires a formatted report with specific fields populated, the gap between those two things is where manual rework enters the process. Every gap between AI output and workflow requirement is a potential source of AI Efficiency Tax, and the map should make every one of those gaps visible.

Step four is to identify where efficiency is being gained or lost. For each AI touchpoint on the map, ask whether AI is making that step faster, slower, or the same. Ask whether AI is creating work for the next step or reducing it. Ask whether AI is producing consistent output across all the people who use it at that step, or whether output quality varies depending on who runs the tool, how they prompt it, and what context they provide. Ask whether the workflow moves more smoothly after the AI step than it did before AI was introduced, or whether the AI step has created a new friction point that didn't exist in the pre-AI version of the process. The next section of this chapter walks through seven specific signals to read the map against, and those seven signals are the diagnostic framework that turns the questions above into findings the leadership team can act on.

The answers to those questions are the findings that make the map useful. Write them directly onto the map, next to the relevant step, in plain language. "Review takes 45 minutes per output." "AI output requires manual reformatting before it can be used in step 6." "Three different employees use three different prompts at this step, producing inconsistent outputs that require reconciliation before delivery." "Exception handling at step 4 adds an average of two hours to the cycle when it occurs." Those specific, written findings are what turn the Workflow Reality Map from a picture of the workflow into a decision tool for improving it.

Reading the Map. What to Look for Once It's Built

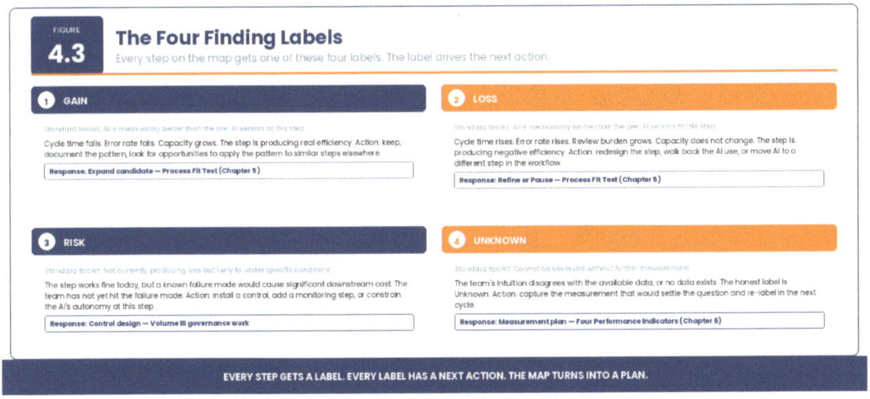

A completed Workflow Reality Map tells a story. The job of leadership is to read that story accurately and then act on what it says. Most workflow maps, when built honestly, reveal that AI is doing three different things simultaneously in different parts of the process: helping in some places, neutral in others, and quietly making the workflow more expensive in a few specific spots that nobody had identified before the map was built.

The first signal to look for is AI steps that add review burden rather than reduce it. This shows up on the map as an AI touchpoint followed by a review step that is longer, more complex, or involves a more senior person than the review step that existed before AI was introduced. If AI is generating output that requires a manager to spend forty-five minutes checking before it can move forward, and the pre-AI version of that step required a team member to spend thirty minutes producing the same output without a separate review stage, the AI step has added cost rather than removing it. The tool made the production faster and the total cycle slower. That's a review burden finding, and it points directly to one of two fixes: either the AI output standards need to be tightened so the review can be shortened, or the review step needs to be redesigned so it happens earlier in the process before cost compounds.

The second signal is AI outputs that trigger rework. This appears on the map as a correction loop after an AI touchpoint, where output goes back for revision before it can continue through the workflow. A single correction loop is a finding. Multiple correction loops at the same step, or correction loops that appear consistently across different team members using the same tool, are a pattern. Patterns point to systemic causes, not individual errors. If the same AI step is triggering rework in one out of every three cycles, the problem isn't a bad prompt on a bad day. The problem is that the AI tool is being given insufficient input, unclear instructions, or context it can't access, and the workflow needs to be redesigned to provide what the tool needs before it runs rather than correcting what it produces after it runs.

The third signal is workflow branches that have become shadow processes. These show up on the map as steps that happen in parallel to the official process, often running through tools, communication channels, or approval paths that don't appear in any documentation. Shadow processes are important findings because they reveal that the official workflow has failed in a specific way. The team built a workaround because the approved process didn't work, didn't work fast enough, or didn't produce results the team trusted. When AI is present in the official workflow but the team is routing work around it through a shadow process, the shadow process is telling leadership something direct: the AI step isn't working the way it was intended, and the team has already figured that out even if leadership hasn't.

The fourth signal is handoffs where human and AI work don't connect cleanly. This appears on the map as a gap between what AI produces at one step and what the next step requires as its input. The gap might be a format mismatch. The AI produces a narrative summary, but the next step requires a structured data table. The gap might be a content mismatch. The AI produces a general draft, but the next step requires client-specific details that the AI didn't have access to. The gap might be a timing mismatch. The AI produces output faster than the next step can consume it, creating a queue that accumulates and introduces delay. Every handoff gap has a specific cause and a specific fix. The map makes the gap visible. The fix requires understanding what the next step actually needs and redesigning the AI step to produce it.

The fifth signal is stages where the process still slows down even though AI is involved. This is the most important finding the map can produce, because it reveals where the true bottleneck lives. If AI is active at step three but the process consistently slows at step five, the bottleneck isn't at step three. It's at step five. And whatever is causing the slowdown at step five, whether it's a capacity constraint, an approval delay, a system limitation, or a workload imbalance, AI at step three can't fix it. Speeding up step three only means that work arrives at step five faster, where it waits longer. The map shows this pattern clearly, and it's one of the most valuable things it can reveal, because without it, leadership might invest further in optimizing step three while the real problem at step five continues to constrain the entire workflow.

The sixth signal is output inconsistency across team members using the same AI tool. This shows up on the map as the same workflow step producing visibly different outputs depending on who runs the AI at that step. One analyst's AI-generated proposal looks different from another analyst's. One team member's AI-assisted report includes analysis that another team member's version omits entirely. The map should record this directly. Note who runs the AI step, note the variation in what each person produces, and quantify the downstream reconciliation work that variation creates. Output inconsistency is the third Tax component from Chapter 3 made visible at the workflow step where it originates.

The fix is the standardization work covered later in this chapter and developed in full in Chapter 10.

The seventh signal is management overhead concentrated around specific AI touchpoints. This shows up on the map as steps where managers are spending unscheduled time answering questions, resolving disputes about whether the AI output is acceptable, or making judgment calls about how to handle exceptions the workflow never anticipated. The map should note which AI touchpoints generate the most management overhead per week and which managers are absorbing it. Management overhead is the fourth Tax component from Chapter 3, and the Reality Map is the instrument that locates where in the workflow that overhead is being produced. The fix is usually a combination of clearer standards (so fewer questions reach the manager), explicit ownership assignment at the AI step (so the manager isn't the default decision-maker), and the workflow corrections covered in the next section of this chapter.

Each signal the map surfaces is tied directly to an optimization opportunity. The map isn't a picture to admire. It's a decision tool. Every finding on the map should produce a specific question: what does this finding require us to change? That question is what moves the map from documentation to action, and it's what the next section is built to answer.

Turning Map Findings Into Optimization Decisions

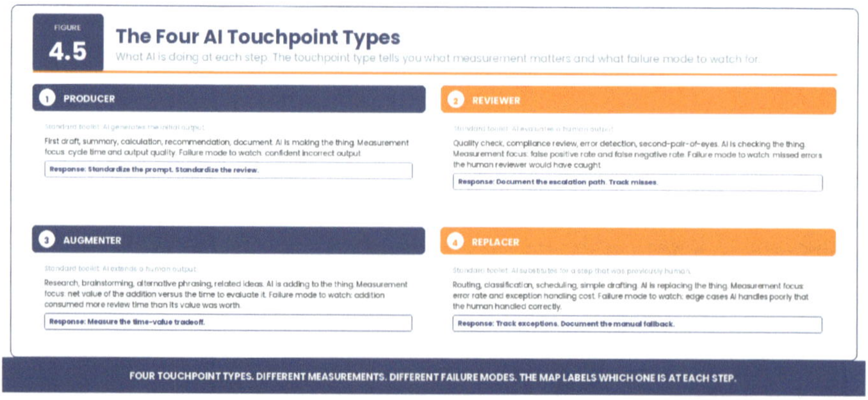

A Workflow Reality Map is finished when leadership decides what the map requires them to change. Not when the workflow is drawn. Not when the findings are labeled. When a decision has been made for each finding that tells the business what to do next. That decision is the output that makes the mapping work valuable, and it's the input that feeds directly into the AI Process Fit Test in the next chapter.

The way to convert each finding into a decision is to run it through a short diagnostic sequence. Ask what type of problem the finding represents. Is it a standards problem, meaning the AI tool is producing inconsistent or low-quality output because there are no shared guidelines for how to use it? Is it a placement problem, meaning AI has been inserted at the wrong point in the workflow where it can't access the inputs it needs or where its output doesn't match what the next step requires? Is it a process design problem, meaning the workflow itself is broken in a way that AI can't fix and may actually be accelerating? Is it an ownership problem, meaning nobody has been clearly assigned responsibility for the quality and consistency of AI output at that step? Or is it a shadow process problem, meaning the official workflow has been bypassed because it doesn't work, and the real process needs to be redesigned around how the work actually gets done?

Each type of problem has a corresponding decision.

A standards problem requires standardization. This means creating shared prompting guidelines, defining output quality criteria, and establishing a minimum standard that every team member using the AI tool at that step must meet before the output moves forward. Standardization is the fastest and least disruptive fix available, and it often reduces review burden significantly within the first thirty days of implementation. The decision is specific: who will write the standards, what format they'll take, when they'll be reviewed with the team, and how compliance will be confirmed.

A placement problem requires repositioning. The AI tool needs to move to a different point in the workflow where it has access to the inputs it needs and where its output connects cleanly to the next step's requirements. This is a more significant change than standardization because it affects the sequence of the workflow, not just the standards within a step. The decision needs to include what the new placement looks like, what inputs AI will receive at the new position, who is responsible for providing those inputs, and what the handoff to the next step will look like after the repositioning.

A process design problem requires redesign before any further AI optimization. If the workflow is fundamentally broken in a way that AI is accelerating rather than fixing, adding more AI investment into that workflow will compound the problem. The decision here is to pause AI optimization for that specific workflow, fix the underlying process issue first, and then re-evaluate AI placement once the process is stable. This is not a failure. It's a correct sequencing decision. AI cannot fix a broken process. It can only run faster inside one (Bishop, 2021), which makes the break more expensive.

An ownership problem requires assignment. Someone needs to be named as the owner of AI output quality at that step. The owner is responsible for monitoring output consistency, reviewing the standards, flagging problems when they appear, and reporting performance metrics to leadership. Without a named owner, standards degrade, shadow processes grow, and review burden accumulates without anyone being accountable for reducing it. The ownership decision is simple but non-negotiable: one person, one workflow step, clear responsibility, and a specific reporting cadence.

A shadow process problem requires a redesign decision at the leadership level. The shadow process exists because the official workflow failed. Leadership has two choices. Either fix the official workflow so it actually works the way it was intended and the team has no reason to route around it, or formally acknowledge that the shadow process is the better process and redesign the official workflow around how the work actually gets done. Both choices are valid. The wrong choice is to ignore the shadow process and keep optimizing the official workflow while the real work continues to move through a path that leadership doesn't govern, doesn't measure, and can't improve.

To make this practical, consider a hypothetical scenario that illustrates how these decisions come together. Return to the 60-person professional services firm. Its COO built a Workflow Reality Map for the client onboarding process, where AI had been introduced to generate initial compliance document drafts. The map revealed four findings. First, the AI step was producing drafts that required an average of 55 minutes of senior review per client, up from 35 minutes of direct drafting before AI was introduced. Second, the AI tool was being prompted differently by three different analysts, producing inconsistent draft quality that the compliance officer had to reconcile before any document moved forward. Third, a shadow process had emerged where two of the analysts were skipping the AI step entirely for a specific document type because the AI output was consistently wrong for that category. Fourth, the true bottleneck in the onboarding process was a manual data entry step at stage seven, not the document drafting step at stage three where AI was focused.

Four findings. Four decisions. The review burden at stage three pointed to a standards problem: shared prompting guidelines needed to be written and enforced. The output inconsistency pointed to the same standards problem and confirmed the priority. The shadow process pointed to a use case removal decision for that specific document type: AI shouldn't be used there until the underlying accuracy issue is resolved. And the bottleneck finding at stage seven pointed to a placement decision: the next AI investment should go to stage seven, not to further optimization of stage three. That COO walked out of the mapping exercise with a short, specific decision list that changed where the business invested its next optimization effort, removed a use case that was generating more cost than value, and gave the compliance team a standards framework that reduced review time within six weeks.

Six Recurring Workflow Patterns That Show Up Across SMB AI Operations

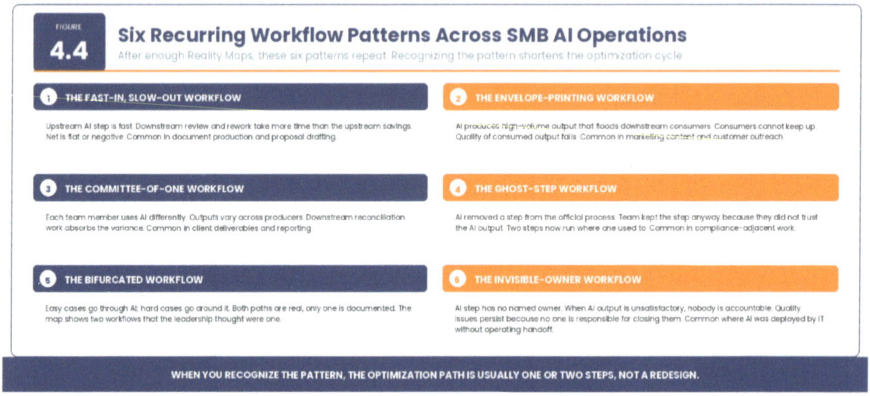

The seven signals above describe what to look for at individual AI touchpoints on the map. The six patterns below describe how those signals combine into recurring presentations that surface across SMB AI operations. The distinction matters because a signal is an atomic finding at one step, while a pattern is a compound finding across multiple steps, often with a specific causal mechanism that points to a specific corrective action. The specific workflows differ across engagements. The structural patterns repeat. Six patterns appear with the highest frequency and are worth naming directly, because the team that recognizes a pattern can move from observation to action without rebuilding the analysis each time.

The first pattern is the bottleneck shift. The designed workflow has AI at the original bottleneck step, the AI is producing output faster than the previous bottleneck did, and the workflow throughput has not improved because the bottleneck moved to a different step the team did not see in advance (Goderis et al., 2005). The Reality Map reveals the new bottleneck downstream of the AI step. The corrective action is workflow redesign at the new bottleneck, not further optimization at the AI step. Continuing to invest in the AI step when the bottleneck has moved produces no throughput gain and adds to the AI Efficiency Tax.

The second pattern is the review-loop loop. AI generates output, a reviewer corrects it, the team uses the corrected output, and the reviewer's correction notes are not feeding back into the AI prompt or the workflow standards. The same corrections recur week after week, producing review burden that never declines and rework cost that compounds. The Reality Map reveals the absence of the feedback loop between reviewer corrections and upstream prompting. The corrective action is the standardization work in Chapter 10, applied to the reviewer correction patterns, with the corrections captured in updated prompts or workflow standards within thirty days of detection.

The third pattern is the shadow channel, which the Library has already encountered in different forms. Volume II Chapter 7 treats the shadow channel as a friction signal during the readiness diagnostic. The Reality Map in this chapter surfaces the same shadow channel as one of five workflow signals read inline during post-deployment mapping. The diagnostic instrument changes; the underlying pattern is the same. The designed workflow routes work through a specific AI tool and review process. The real workflow has work routed through a different tool the team chose informally, often a personal subscription, with a different review process that nobody documented. The Reality Map reveals the shadow channel handling a meaningful share of the work, sometimes the majority. The corrective action depends on whether the shadow channel is producing better results or worse. Better: bring it into the approved environment with proper governance. Worse: identify why the team chose it and address the underlying need before redirecting the work back to the approved channel.

The fourth pattern is the workaround steady state. The designed workflow assumes AI output is usable as-is. The real workflow includes a quiet manual step in which a team member rewrites or re-formats AI output before it enters the rest of the workflow. The workaround was developed informally during the first weeks of AI deployment, became a standard part of the team's routine, and is no longer recognized as a workaround. The Reality Map reveals it for what it is.

The corrective action is to address whatever is causing the AI output to require the rewrite (often a prompt quality issue), then remove the workaround once the output meets the standard.

The fifth pattern is the dual-version race. The team has been producing both an AI-drafted version and a manually-drafted version of the same output, and uses whichever one looks better at the moment of delivery. The dual-version race is most common in client-facing work where the team has lost trust in the AI output and is hedging by producing both. The Reality Map reveals it through the volume mismatch between AI output and final delivered work. The corrective action is to address the underlying trust gap, either by improving AI output quality to the point that the manual version is unnecessary, or by acknowledging that the workflow is not a good AI fit and using the Process Fit Test from Chapter 5 to make the formal decision to pause or remove.

The sixth pattern is the silent abandonment. The designed workflow assumes AI is in regular use. The real workflow shows AI being used early in the team's adoption period, then gradually used less, then effectively abandoned, while AI usage continues to appear in the dashboard because someone is still logging in occasionally to maintain appearances. The Reality Map reveals the gap between reported usage and actual usage. The corrective action depends on why the team abandoned the AI. Usually it is because the AI was not producing usable output, and the team's abandonment is the correct response. The leadership decision is whether to formally remove the use case or invest in making AI work for this workflow.

Recognizing one of these six patterns early in the Reality Map work is one of the most useful shortcuts the discipline offers. The patterns are not exhaustive and new patterns will appear over time. But the six above cover the most common SMB AI presentations, and a team that has the patterns named in advance moves faster from observation to action than a team that is encountering each pattern for the first time.

When the Reality Map Reveals One Workflow Disguised as Several

Most workflows in a business stand alone. They have distinct teams, distinct inputs, distinct outputs, and distinct review processes. Occasionally the Reality Map work reveals a different pattern: what the business has been treating as several distinct workflows is actually one workflow with multiple presentations. This is the merged workflow pattern. It is less common than the typical case where workflows are correctly identified as separate, and it is more consequential when it appears, because the optimization work has been splitting effort across what should have been a single unified intervention. The Reality Map for each workflow looks similar to the Reality Maps for the others, the same team members appear in each one, and the inputs and outputs partly overlap.

The 60-person professional services firm carried through this Library encountered the pattern in its second quarter of Volume IV work. The firm had been treating proposal drafting, project status reporting, and client research summarization as three separate workflows with separate AI use cases, separate Scorecards, and separate ROI calculations. The Reality Map work in Chapter 4 revealed that the same three team members produced output for all three workflows, that the source documents fed into all three workflows from the same set of internal repositories, and that the review process for all three was effectively the same. The three workflows were three presentations of one workflow.

Once the firm recognized the pattern, the optimization work simplified substantially. Instead of three Scorecards, one Scorecard covered the combined workflow with three output channels. Instead of three Reality Maps, one Reality Map covered the workflow with the three channels noted as variations. Instead of three ROI calculations, one calculation covered the combined activity with the three output types weighted by volume. The simplification saved meaningful time in the next quarterly review cycle and produced a clearer picture of where the workflow was actually performing well versus where the optimization effort should focus.

The diagnostic for the merged workflow pattern is straightforward. If two or more workflows being treated separately share the same team members, the same source inputs, and the same review process, they may be one workflow with separate presentations. The Reality Map will surface the overlap if it is drawn with the team's actual production activity in mind rather than the designed boundaries. The corrective action is to consolidate the optimization work onto the combined workflow, with the variations noted, and to manage the Scorecard, ROI, and Brief at the combined level rather than the variation level. Most leadership teams find the consolidated view easier to manage and more useful for decision-making than the fragmented view it replaces.

That's what the Workflow Reality Map produces when it's built honestly and read carefully. Not a picture of the workflow. A decision list that tells leadership exactly what to fix, in what order, and why.

Putting It Into Practice

Chapter 3 produced the Tax estimate. This chapter built the Reality Map methodology. The two outputs combine into a single piece of work the reader can do now: walk the highest-Tax workflow as it actually runs, label every finding, classify every AI touchpoint, and hand the resulting map to Chapter 5 as the input to the Fit Test. Before turning the page, complete this work on the workflow Chapter 3 ranked highest.

First, walk the workflow as it actually runs. Spend a representative week observing the work, not a peak week and not a slow week. Document every step the work passes through, including the informal approvals, manual workarounds, exception handling, and shadow steps that consume real time but never appear on the official process diagram. The designed workflow is not the subject. The running workflow is. Two or three team members who do the work daily, not the manager who designed it, are the right observation targets. The map is only as honest as the people who produced it.

Second, label each finding. For every step on the map, attach one of four labels. Gain marks a step where AI is producing real efficiency, supported by observable evidence rather than assumption. Loss marks a step where AI is producing Tax, traceable to one of the five components from Chapter 3. Risk marks a step where data exposure, compliance friction, or quality failure is possible even if no Tax is being generated yet. Unknown marks a step where the evidence is insufficient to decide, which becomes a follow-up observation target rather than a default-positive assumption.

Third, classify each AI touchpoint by type. Input generation, output review, decision support, and autonomous action carry different risk profiles and different optimization moves. A workflow with three input-generation touchpoints behaves differently from a workflow with one autonomous-action touchpoint, even if the Tax totals look similar. Chapter 5 will use this classification to score each use case across the five Fit Test dimensions, and the touchpoint type is one of the inputs.

Fourth, hand the completed map to the Fit Test. The map is the input. The Fit Test is the instrument that converts the input into an explicit decision: expand the use case, refine it, pause it, or remove it. Each decision is paired with a named owner and a date. A map without a decision is a document. A map with a decision attached is a tool. The Fit Test is what attaches the decision, and the next chapter walks through how to run it.

The map is built. The findings are labeled. The decisions are typed. The work of looking at a workflow as it actually runs, not as it was designed to run, has produced what the next several chapters depend on: a picture of where AI belongs, where it does not, and where the AI Efficiency Tax is entering the process step by step. The next chapter takes that picture and runs the AI Process Fit Test against it, producing the explicit expand-refine-pause-remove decision for each active use case.

FIGURE 4.6

Putting It Into Practice

Build the Reality Map for the workflow Chapter 3 ranked highest by Tax. Convert the map into a decision input the Fit Test can act on

THE FOUR INPUTS YOU NEED BEFORE WALKING THE WORKFLOW

I1 — The workflow itself
The highest-Tax workflow from the Chapter 3 ranking, named by what it produces, not by the AI tool inside it. The map is built around the work, not the software

I2 — The people who actually run it
Two or three team members who do the work daily, not the manager who designed it. The designed workflow and the real workflow diverge in their hands, and that gap is what the map captures

I3 — One representative week of observation
Not a normal week, not a peak week, a representative one. The findings are valid only if the workflow being mapped is the workflow that actually runs most of the time

I4 — The Chapter 3 Tax estimate as reference
The five-component Tax figure for this workflow from Chapter 3. The map will identify where each component is entering the process, so the Tax estimate is the comparison anchor

1 WALK THE WORKFLOW AS IT ACTUALLY RUNS

Document every step the work moves through, not every step the SOP describes. Capture handoffs, informal approvals, manual workarounds, exception handling, and the shadow steps that consume time every week. The designed workflow is not the subject. The running workflow is

Owner: Workflow Owner with the two or three team members

2 LABEL EACH FINDING

For every step, attach one of four labels: gain (AI is producing real efficiency here), loss (AI is producing Tax here), risk (data exposure, compliance, or quality risk), or unknown (insufficient evidence yet to decide). The labels are what convert observations into decisions.

Owner: AI Efficiency Lead — within two days of completing the walk

3 IDENTIFY EACH AI TOUCHPOINT TYPE

For every step where AI is involved, classify the touchpoint: input generation, output review, decision support, or autonomous action. Different touchpoint types carry different risks and different optimization moves. Chapter 5 will use this classification.

Owner: AI Efficiency Lead with Workflow Owner

4 HAND OFF TO THE FIT TEST

The completed map (steps, labels, touchpoint types, Tax locations) is the input to Chapter 5's AI Process Fit Test. The Fit Test produces the explicit expand-refine-pause-remove decision per use case. A map without a decision is a document. A map with the Fit Test applied is a tool.

Owner: Executive Sponsor schedules Fit Test session

A MAP WITHOUT A DECISION IS A DOCUMENT. A MAP WITH A DECISION ATTACHED IS A TOOL.

Chapter 5: The AI Process Fit Test

Deciding Which Workflows AI Should Actually Touch

CHAPTER 5 OUTCOMES

At the end of this chapter, you will be able to:

1. Define the AI Process Fit Test and explain how it differs from the Reality Map and the Tax Calculator.
2. Score a workflow on the five dimensions of fit: Workflow, Data, Team, Financial, and Risk.
3. Produce one of four operating decisions per use case: Expand, Refine, Pause, or Remove.
4. Run a scoring session with the workflow owner and team members in ninety minutes with documented evidence per dimension.
5. Walk back an AI use case without losing leadership credibility — the structured language for the Pause and Remove decisions.

THE FIT TEST CONVERTS THE QUESTION 'IS THIS WORKING?' INTO A DEFENSIBLE OPERATING DECISION.

The Problem This Chapter Solves: Not Every Workflow Is Ready for AI

The Workflow Reality Map from Chapter 4 gives you something most leadership teams have never had: an honest picture of what's actually happening inside your AI-supported workflows. You can see the real steps, the real handoffs, the real review burden, and the real cost. That visibility is essential. But visibility alone doesn't tell you what to do next. It tells you what exists. The AI Process Fit Test tells you what belongs.

That distinction matters more than it might seem at first.

Seeing that a workflow has a rework problem doesn't automatically mean AI should stay or go. Seeing that review burden is high doesn't automatically mean the use case is wrong. A workflow can have real problems and still be a strong candidate for AI once those problems are corrected. Conversely, a workflow can look clean on the surface and still be a fundamentally poor fit for AI because it lacks the underlying conditions AI needs to produce consistent, measurable improvement. The Workflow Reality Map shows you the current state. The AI Process Fit Test tells you whether AI belongs in that state, or in any improved version of it.

One of the most expensive AI mistakes a business can make is placing AI inside a workflow that was never stable, measurable, or standardized enough to support it. AI doesn't repair a broken process on its own. It runs inside the process as it exists, and if that process is unstable, inconsistent, or poorly defined, AI accelerates the confusion, multiplies the exceptions, and compounds the rework that was already present. The business ends up paying for AI tools and absorbing the cost of a faster version of the same problem it had before the tool was introduced.

The AI Process Fit Test gives you a structured way to avoid that outcome. It evaluates any workflow across five specific dimensions and produces one of four operating decisions: expand, refine, pause, or remove. It protects leadership from treating AI enthusiasm as operational evidence. It forces a more disciplined question before any AI use case is expanded, continued, or approved: does this workflow have the conditions required for AI to create measurable improvement? If the answer is yes, the business knows what to do next. If the answer is no, the business knows that too, and it knows why, which is the information needed to either fix the workflow or redirect the investment.

By the end of this chapter, you'll be able to separate workflows that can support AI from workflows that need repair first, and you'll have a clear, documented decision for each active AI use case in your operation.

One operating distinction matters before the chapter goes further. Volume II Chapter 9 introduced the Expand, Refine, or Pause Decision Protocol, which gives leadership an initial decision about each AI use case based on readiness scoring before broader deployment. The Process Fit Test in this Volume is not a replacement for that decision. It is the recurring post-deployment re-decision that runs every quarter once AI is operational. Volume II asks whether a use case is ready for deployment. Volume IV asks whether a use case that has been deployed for a quarter or more is still earning its operating cost, and if not, what corrective action is required. The decision outputs overlap intentionally: expand, refine, and pause are common to both. The Volume IV addition is remove, which becomes the right decision when a use case that was approved for deployment has run for one or more quarters and the measurement evidence shows it is not producing value the business can defend. Volume II's decision sets the initial trajectory. Volume IV's decision is how the leadership team keeps that trajectory honest as the operating evidence accumulates.

The Five Dimensions of the AI Process Fit Test

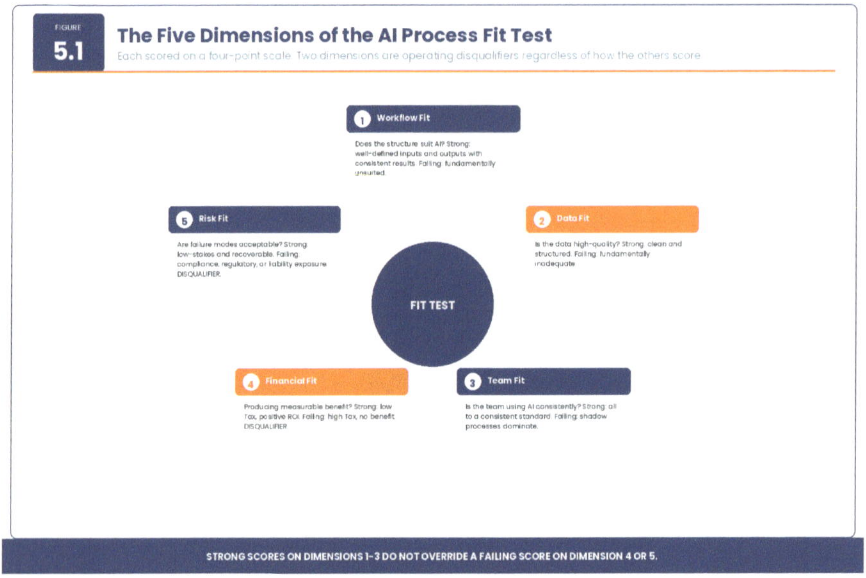

The AI Process Fit Test evaluates a workflow across five practical dimensions. Each dimension asks a specific question about the workflow's current condition. Each question has a scoring range. And the total score across all five dimensions produces a clear picture of whether the workflow is ready for AI, and if not, what's making it unready.

The five dimensions are process stability, output measurability, error tolerance, handoff clarity, and standardization level. Each one is independent, which means a workflow can score well on three dimensions and poorly on two, and that profile tells you something specific about which problems to fix before expanding AI involvement. A low score doesn't mean the workflow is broken. It means the workflow isn't ready for AI yet, and the score tells you exactly why.

Score each dimension on a scale from one to four. A score of one means the condition is absent or severely inadequate. A score of two means the condition exists in a limited or inconsistent way. A score of three means the condition is present and reasonably reliable. A score of four means the condition is strong, documented, and consistently applied. The total possible score across all five dimensions is twenty. Use the score ranges below to interpret the result and assign a decision. Scores from seventeen to twenty support an expand decision. Scores from twelve to sixteen support a refine decision. Scores from seven to eleven support a pause decision. Scores from five to six support a remove decision. Any single dimension scoring a one should be treated as a flag regardless of the total score, because a critical weakness in one area can undermine strong performance in all the others.

Work through each dimension using the findings from your Workflow Reality Map. The map gives you the evidence. The fit test gives you the framework for interpreting that evidence into a decision. These two tools are designed to work together, which is why the fit test comes immediately after the mapping chapter. The map produces the inputs. The fit test converts those inputs into an operating choice.

Dimension one is process stability. This dimension asks whether the workflow follows a predictable pattern or changes constantly. A stable process runs the same way in most cycles. The steps are consistent. The sequence is reliable. The inputs that enter the process are reasonably predictable. The people involved know what to expect at each stage. A score of four means the workflow runs the same way in more than ninety percent of cycles with documented steps and clear ownership at every stage. A score of three means it's mostly consistent with occasional exceptions that are handled in a known way. A score of two means the workflow varies significantly depending on who's running it, what client or project is involved, or what day of the week it is. A score of one means the workflow is essentially unpredictable, with different steps, different sequences, and different people involved in almost every cycle.

AI performs best in stable processes. (Wehenkel et al., 1989) When the workflow changes constantly, AI can't learn from it, can't be prompted consistently for it, and can't produce reliable output across cycles. An unstable workflow that runs AI is likely to produce inconsistent, unreliable outputs that require heavy review and frequent correction, which means the AI Efficiency Tax is almost guaranteed to be high regardless of which tool is used or how well it's implemented.

Dimension two is output measurability. This dimension asks whether the business can define what good looks like for the output this workflow produces. A score of four means there are documented quality criteria for the workflow's output, those criteria are applied consistently, and the business tracks whether outputs meet the standard in every cycle. A score of three means quality criteria exist informally and most people who work in the workflow understand what a good output looks like, even if it's not formally written down. A score of two means quality is evaluated subjectively, with different reviewers applying different standards depending on their own judgment. A score of one means there's no defined standard for what a good output looks like, and quality is assessed entirely by feel.

Output measurability is critical for AI because AI can't produce consistently good output if nobody has defined what good means. (Wang, 2019) Without a measurable standard, review is arbitrary, correction is subjective, and the business has no way to determine whether AI is improving output quality or degrading it. The score for this dimension often reveals why review burden is high: when quality criteria are unclear, reviewers compensate by checking everything, which drives review time up (Blank, 2006) and makes the review step dependent on senior judgment rather than documented standards.

Dimension three is error tolerance. This dimension asks how much risk the workflow can absorb if AI produces a flawed result. A score of four means errors in this workflow are caught before they cause any downstream harm, the correction cost is low, and the business can recover quickly from an AI mistake without significant financial, operational, or reputational consequence. A score of three means errors are generally caught before they reach an external stakeholder, though they may create internal rework or delay. A score of two means errors occasionally reach clients, partners, or other external parties, and when they do, the cost of correction is meaningful. A score of one means errors in this workflow carry significant risk of financial loss, regulatory exposure, client damage, or reputational harm, and the business has limited ability to catch and correct them before they cause impact.

Error tolerance doesn't determine whether AI can be used in a workflow. It determines how much quality control infrastructure needs to surround the AI step before it's safe to run at scale. A low error tolerance score isn't a reason to remove AI. It's a reason to build stronger review checkpoints, tighter output standards, and clearer escalation paths before expanding AI involvement. Ignoring a low error tolerance score and expanding AI anyway is how a business turns a manageable rework problem into a client-facing quality failure or a compliance exposure.

Dimension four is handoff clarity. This dimension asks whether responsibility between people, systems, and AI is clearly defined at every stage of the workflow. A score of four means every handoff in the workflow has a named owner, a defined trigger for when the handoff occurs, and a documented standard for what the receiving party needs from the sending party. A score of three means most handoffs are clear, with occasional ambiguity that the team resolves informally. A score of two means handoffs are frequently unclear, with work sitting in undefined states between steps while team members figure out who's responsible for moving it forward. A score of one means handoffs are consistently unclear, with work frequently stalling, being duplicated, or falling through the gaps between people and systems.

Poor handoff clarity is one of the most common causes of high AI Efficiency Tax. (Agha-Mir-Salim et al., 2025) When AI produces an output but nobody has a clear responsibility for what to do with it next, the output sits in a queue. Or it gets reviewed by multiple people applying different standards. Or it gets corrected by someone who wasn't supposed to touch it and then corrected again by the person who was. Every one of those outcomes adds cost and time to the workflow without adding value. AI can't fix unclear handoffs. It can only produce output faster, which means unclear handoffs accumulate faster when AI is involved.

Dimension five is standardization level. This dimension asks whether the process runs consistently enough for AI to improve it. A score of four means the workflow is fully standardized: shared tools, shared prompts, shared quality criteria, shared review standards, and consistent execution across every team member who participates. A score of three means most elements are standardized, with some variation in how individual team members execute specific steps. A score of two means standardization is partial, with significant variation in tools, prompts, or quality criteria across the team. A score of one means there's essentially no standardization, with every team member running the workflow in a different way using different tools, different inputs, and different standards for what an acceptable output looks like.

Standardization is the dimension that most directly predicts whether AI will produce consistent results (Haque, 2025) across the team or wildly variable ones. A workflow with a standardization score of one will produce AI outputs that differ dramatically depending on who runs the tool (Ranschaert et al., 2021) , how they prompt it, and what context they provide. That variability creates inconsistency cost, management overhead, and review burden that compounds every week. Raising the standardization score before expanding AI involvement is almost always the highest-return improvement a business can make in a low-scoring workflow.

Running the Test: A Practical Scoring Session for a Lean Leadership Team

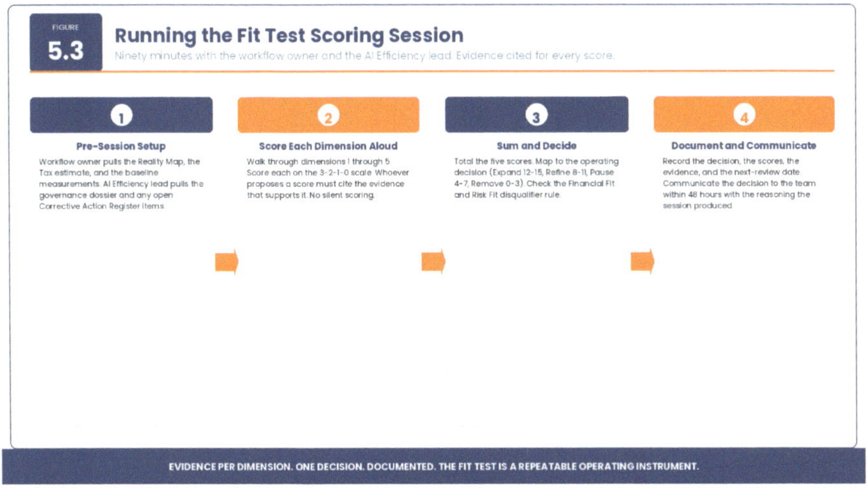

The AI Process Fit Test is designed to be completed by a small leadership group in under two hours per workflow. You don't need a consultant, a data team, or a specialized platform. You need the findings from your Workflow Reality Map, the five scoring criteria above, and honest input from the people closest to the work.

Start by selecting the workflow you're testing. Use your Workflow Reality Map findings as the primary evidence base. The map already contains most of what you need to score each dimension accurately. You're not starting from scratch. You're interpreting existing evidence through a structured scoring framework.

Assemble a small scoring group of two to four people. This should include at least one person who works inside the workflow daily, one person who manages or oversees it, and one person with a financial or operational perspective, such as a CFO, COO, or senior operating leader. Three people with different vantage points on the same workflow will produce a more accurate score than one person completing the assessment alone, because each dimension looks different depending on where you sit relative to the work.

Score each dimension independently before discussing it as a group. Ask each person to assign a score from one to four for each dimension based on their direct knowledge of the workflow. Then compare scores. Where scores align closely, the evidence is probably clear. Where scores diverge significantly, the divergence itself is a finding. A manager who scores handoff clarity at three and a team member who scores it at one are telling you something important: the handoffs look clearer from above than they feel from inside the workflow. That gap is worth understanding before making an expansion decision.

After scoring all five dimensions, add the scores and record the total. Then review any dimension that scored a one separately from the total, because a single critical weakness can make a high total score misleading. A workflow that scores four on four dimensions and one on error tolerance is not a strong AI candidate if the error tolerance problem exposes the business to client-facing quality failures or regulatory risk. The one-score flag overrides the aggregate total in those cases.

Document the scores, the rationale for each score, and any significant disagreements within the scoring group. That documentation is the record that justifies the decision that follows. It's also the baseline against which the next fit test will be compared, which is how the business tracks whether workflow conditions are improving over time.

To make this concrete, consider a hypothetical scenario. Return to the 40-person accounting firm. Its operations manager is responsible for a return intake workflow where AI was introduced six months ago to generate initial review summaries from client-submitted documents. She runs the AI Process Fit Test with her team using the Workflow Reality Map findings from the previous week. Process stability scores a three: the workflow is mostly consistent but has two common exception types that each follow different paths. Output measurability scores a two: the team has an informal sense of what a good summary looks like, but there are no written quality criteria and different reviewers apply different standards. Error tolerance scores a two: errors in the summary occasionally reach the reviewing partner without being caught, and when they do, they add an average of four hours to the return preparation cycle. Handoff clarity scores a three: most handoffs are clear, but the step between AI output and partner review has an ownership gap that causes delays when the primary reviewer is unavailable. Standardization scores a two: three different analysts use three different prompt approaches, producing summaries of noticeably different quality and depth. Total score: twelve. Decision: refine. The use case has potential, but output measurability, standardization, and error tolerance need to improve before AI involvement should be expanded. That's a specific, actionable result from a two-hour session, and it tells the operations manager exactly which problems to fix and in what order.

Interpreting Your Fit Test Results: Expand, Refine, Pause, or Remove

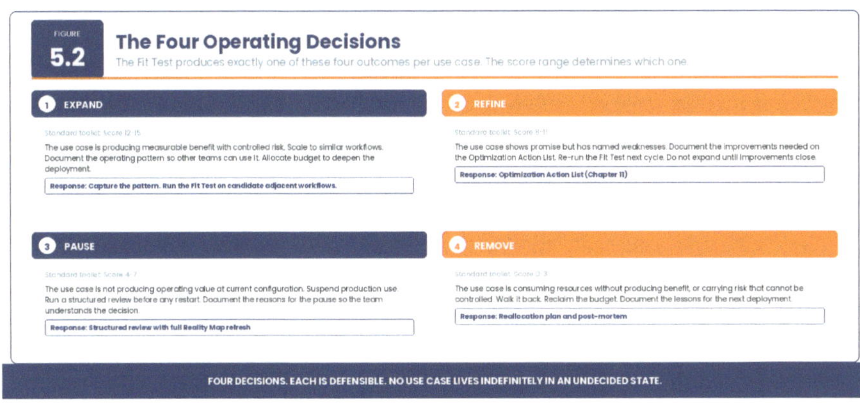

Once the workflow is scored, the AI Process Fit Test produces one of four operating decisions. Each decision has a specific meaning, a specific set of next actions, and a specific communication approach. The decision isn't a judgment about the team or the tool. It's an operating assessment of whether the workflow conditions support AI at the current level of involvement, and what the business should do next based on those conditions.

An expand decision means the workflow is stable, measurable, clear, and standardized enough for AI to scale with confidence. The five dimensions are all scoring at three or four. The AI Efficiency Tax is low or declining. The downstream results are measurable and positive. Review burden is manageable and decreasing as the team gains experience with the standards. An expand decision authorizes the business to increase AI involvement in this workflow, whether by processing higher volume, adding additional AI-assisted steps, or applying the same AI approach to similar workflows where the conditions are comparable. The documentation for an expand decision should record the scores, the evidence behind them, and the specific expansion being authorized, so the decision is defensible and the baseline is clear for the next fit test cycle.

A refine decision means the use case has real potential but the workflow, the AI instructions, the review structure, or the output standards need correction before expansion is appropriate. A refine decision is the most common outcome for AI use cases that have been running for three to twelve months without structured measurement. The tool is probably doing something useful. The workflow conditions are probably good enough in some dimensions and weak in others. The refine decision names which dimensions need improvement, assigns ownership for each improvement, and sets a timeline for re-scoring. It doesn't stop AI from operating in the workflow. It stops the business from expanding AI involvement until the underlying conditions improve enough to support it.

The specific actions that follow a refine decision depend on which dimensions scored low. A low process stability score points to workflow documentation and exception handling as the first priority. A low output measurability score points to writing quality criteria and establishing a review standard as the first priority. A low error tolerance score points to redesigning the review checkpoint structure before any further expansion. A low handoff clarity score points to ownership assignment and handoff documentation as the first priority. A low standardization score points to shared prompting guidelines and output standards as the first priority. Each low-scoring dimension has a corresponding fix, and the refine decision should name the fix, the owner, and the timeline explicitly.

A pause decision means AI should temporarily stop touching the workflow until the business stabilizes the process, clarifies ownership, or reduces risk. A pause is appropriate when the total score is in the seven to eleven range or when a single dimension scores a one in an area that creates meaningful risk. A pause is not a failure. It's a risk management decision. It means the business recognized that the workflow conditions aren't safe for AI at the current level of involvement, and it made a responsible choice to stop and fix the foundation before continuing. The pause decision should include a clear description of what needs to change before the workflow is re-evaluated, a timeline for that re-evaluation, and a named owner who is responsible for driving the improvement work and bringing the next fit test results back to leadership.

A remove decision means the AI use case doesn't belong in that workflow. The fit is fundamentally wrong, the risk is too high, or the hidden cost consistently exceeds the value the tool creates. A remove decision is the hardest one to make and the most important one to make correctly. It requires leadership to acknowledge that an AI use case that was approved, implemented, and perhaps championed by someone on the team needs to stop. The remove decision should be documented with the scoring evidence that supports it, the specific reasons the fit is wrong, and a clear statement of what the business is choosing to do instead. That documentation protects leadership, demonstrates operating discipline, and creates a record that shows the CFO, board, or auditor that AI decisions are made on evidence rather than enthusiasm.

The four decisions form a complete operating framework. Every active AI use case in the business should be able to produce one of these four decisions when run through the fit test. If a use case can't be scored because the workflow is too poorly defined to evaluate, that itself is a finding. It means the workflow needs to be mapped and documented before any AI decision can be made about it, which is a version of the pause decision applied at the pre-test stage.

How to Walk Back an AI Use Case Without Losing Credibility

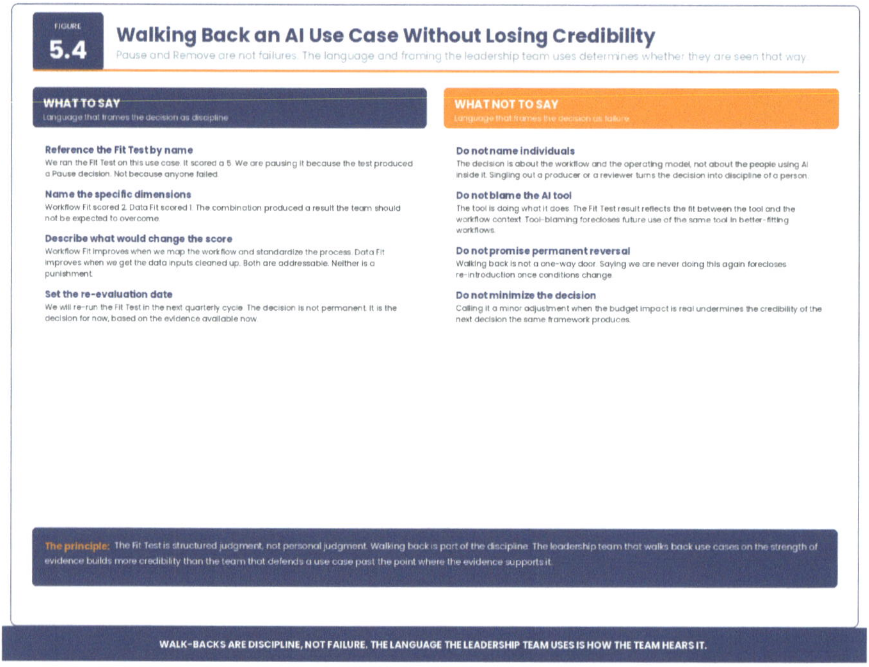

Pausing or removing an AI use case is one of the most useful things a leadership team can do. It's also one of the most uncomfortable. Someone approved the use case. Someone championed it. Someone told the team it was going to improve the workflow. And now the fit test is saying it shouldn't be there, or shouldn't be expanded, or needs to stop until the underlying conditions improve. That's a real organizational moment, and how leadership handles it determines whether the AI governance culture strengthens or quietly fractures.

The first thing to understand is that walking back a use case is not an admission that AI failed. It's evidence that leadership is managing AI with discipline instead of ego. A business that tests its AI use cases, measures the workflow conditions, finds the fit inadequate, and makes a responsible decision to refine, pause, or remove is doing exactly what sound operating discipline requires. That's not a failure story. It's a governance story, and it's a strong one.

The communication framework for a walk-back has three components. The first is the evidence statement: the business tested the use case using a structured scoring process, and the workflow conditions don't currently support AI at the level of involvement that was planned. The second is the decision statement: based on that evidence, the business is choosing to refine, pause, or remove the use case, and here's specifically what that means for the workflow and the team. The third is the forward statement: here's what needs to change before the decision is revisited, who owns that work, and when the next evaluation will happen.

That three-part structure removes ambiguity, prevents the decision from feeling personal, and gives the team a clear picture of what comes next. It also signals that the decision is based on operating evidence rather than political pressure, executive preference, or a general sense that AI isn't working. When people understand why a decision was made and what would need to change for it to be revisited, they're far more likely to engage with the improvement work than if the decision feels arbitrary or top-down.

The person who originally championed the use case deserves specific attention in this conversation. They took a risk by advocating for something new. That risk-taking is valuable, even when the outcome requires a course correction. The message to that person should be direct: the use case was worth testing, the test produced useful information, and the decision to refine or pause is a responsible next step that the evidence supports. Their advocacy got the business to a point where it could learn something real about the workflow. That's a contribution, not a failure.

Documentation is a non-negotiable part of the walk-back process. A well-documented pause or remove decision strengthens the AI operating record and demonstrates to any CFO, board member, lender, or auditor who reviews it that leadership is governing AI decisions honestly. The documentation should include the fit test scores, the specific dimensions that drove the decision, the decision itself, the rationale in plain language, and the next steps with owners and timelines. That record is not just protection. It's evidence of operating discipline, and it belongs in the same AI governance file that Volumes I through III helped the business build.

One practical way to reduce the discomfort of a walk-back is to frame the fit test as a standard operating cadence from the beginning, before any specific use case needs to be walked back. When every AI use case goes through a structured fit test on a regular schedule, a pause or remove decision is just the normal output of a normal process. It doesn't single out any particular use case or any particular champion. It's just what the business does to make sure AI is placed correctly and performing well. That framing, established early and applied consistently, makes walk-backs feel like operating discipline rather than organizational embarrassment.

The AI-First Workflow: Scoring Workflows That Did Not Exist Before AI

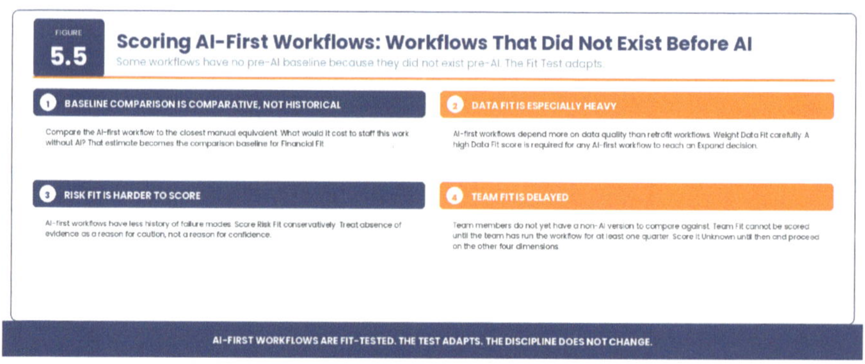

A subset of workflows in any SMB AI operation did not exist before AI. The team did not draft daily competitive intelligence briefs until the AI made it possible. The firm did not produce monthly trend reports for clients until the AI made the production cost feasible. The business did not generate personalized customer follow-ups at this scale until AI removed the labor constraint. These workflows have no pre-AI baseline to measure against because they were born after AI was introduced. The Process Fit Test was designed primarily for workflows where AI replaced or augmented existing manual processes, but it adapts to AI-first workflows with a different scoring frame.

The first adaptation is the baseline question. For an existing workflow, the fit test asks whether AI is producing better, faster, or cheaper outputs than the pre-AI process produced. For an AI-first workflow, the question becomes whether the new workflow is producing value the business would not have without AI. The baseline is the business state before the workflow existed, not before AI was introduced into the workflow. The relevant comparison is whether the resources the AI-first workflow consumes (subscription, review labor, management overhead) are less than the value it produces, not whether AI is faster than the manual alternative that never existed.

The second adaptation is the workflow stability question. The Process Fit Test in its standard form assumes the workflow has been running long enough for the team to understand its inputs, outputs, and exception cases. An AI-first workflow that has been running for two months may not yet have produced enough operating history for the stability question to be answerable. The adaptation is to defer the fit test until the workflow has at least one full quarter of stable operation, treating the first quarter as a setup period in which fit cannot yet be assessed. The team can run the other fit test dimensions on the AI-first workflow but should mark the stability dimension as not-yet-assessable and revisit it after the first stable quarter.

The third adaptation is the value verification question. For existing workflows, the value AI produces is typically measurable against the displaced manual work. For AI-first workflows, the value needs to be verified directly. The team that runs the daily competitive intelligence brief needs to confirm that the brief is actually being read and acted on by the people it was designed for. If the brief is not being read, the workflow is producing volume without value, and the fit test should produce a refined or pause decision regardless of how cheaply the brief is being produced. AI-first workflows have a higher risk of becoming Phantom Productivity at scale because there is no manual benchmark constraining their output (Gates et al., 2026) to a useful volume.

With these three adaptations, the Process Fit Test produces useful decisions for AI-first workflows. The most common outcomes are expand (the AI-first workflow is producing real value and should be extended further), refine (the workflow is producing some value but is not yet at full fit), and remove (the workflow looked promising in concept but is not producing measurable value in operation). Pause is less common for AI-first workflows than for existing workflows, because the cost of pausing is usually higher than the cost of removing entirely.

The Fit Test for Cross-Functional Workflows

Some workflows cross team boundaries. The proposal that originates in business development, gets technical review from operations, and final pricing approval from finance is one workflow with three functional owners. The customer onboarding that runs through sales, customer success, and operations is one workflow with three teams. AI inserted into a cross-functional workflow has fit characteristics that differ from AI inserted into a single-team workflow, and the Process Fit Test requires several specific considerations when applied to the cross-functional case.

The first consideration is ownership. The standard Process Fit Test produces a decision (expand, refine, pause, remove) that becomes the operating instruction for the workflow's owner. A cross-functional workflow may not have a single owner who can act on the decision. The fit test result for a cross-functional workflow needs to be paired with an explicit ownership assignment, often the senior leader who owns the largest functional segment of the workflow. Without the ownership assignment, the fit test result will not produce action, because each functional owner will reasonably assume one of the others is acting on it.

The second consideration is the scoring panel. A standard Process Fit Test is scored by the workflow owner and one or two senior reviewers. A cross-functional fit test should be scored by representatives from each function the workflow crosses, with the scores aggregated rather than averaged. The cross-functional fit test surfaces disagreements between functions about whether the workflow is performing, and those disagreements are often the most useful diagnostic output. A workflow that the finance function scores as fit and the operations function scores as unfit is producing value for one function while costing the other, and the corrective action is to address the cost-bearing function's experience before further expansion is approved.

The third consideration is the dependency mapping. Cross-functional workflows have dependencies that single-team workflows do not. A change to the AI step in business development may produce downstream effects on operations review or finance approval that the business development team does not see. The fit test for a cross-functional workflow should include explicit dependency mapping: which other functions are affected by the AI step, what the downstream effect has been, and whether the affected functions concur with continued AI involvement. The dependency mapping often surfaces operating frictions that the single-function fit test would miss.

The output of a cross-functional fit test is a decision plus an ownership assignment plus a dependency record. That output is more complex than a single-team fit test produces, but the workflow it covers is also more complex, and the additional discipline pays back in better-aligned cross-functional execution. Leadership teams that adopt this protocol for cross-functional workflows usually find that the fit test process becomes a vehicle for cross-functional alignment that the team had been struggling to achieve through other means.

Building a Fit Test Cadence Into Your Operating Rhythm

The AI Process Fit Test is not a one-time approval gate. It's a recurring operating discipline. Workflows change. Teams adapt. Tools evolve. Business conditions shift. A workflow that scored seventeen and earned an expand decision six months ago may look very different today if the team has grown, the client mix has changed, or the AI tool has been updated in ways that affect how it performs inside the process. Fit scores are not permanent. They reflect the workflow conditions at the time of the test, and those conditions need to be rechecked regularly.

The minimum cadence for active AI use cases is quarterly. Every AI-supported workflow that is actively running should go through a fit test re-score at least once every ninety days. That cadence keeps the scores current, surfaces workflow drift before it becomes expensive, and gives leadership a regular opportunity to confirm that expand decisions are still justified and that refine decisions have produced the improvements they were supposed to produce.

Certain events should trigger an immediate re-score outside the quarterly cadence. When a workflow changes materially, meaning new steps are added, ownership changes, or the volume processed through the workflow increases significantly, the fit conditions may have shifted enough to change the decision. When a new AI tool is introduced or an existing tool is significantly updated, the error tolerance and standardization dimensions in particular may need to be re-evaluated. When quality issues appear in a workflow that previously scored well, the output measurability and process stability dimensions should be re-scored before the problem compounds. And when the AI Efficiency Tax for a workflow begins to rise after a period of stability, that's a signal that one or more fit dimensions have deteriorated and the workflow needs a fresh evaluation.

Ownership for each fit test cycle needs to be assigned clearly. Every active AI use case should have a named owner who is responsible for gathering the evidence needed for the re-score, completing the scoring session, documenting the result, and escalating to leadership when the decision recommendation changes. Without a named owner, the cadence will slip. Quarterly re-scores will become semi-annual, then annual, then never. And the business will end up in the same position it was in before the fit test was introduced: running AI use cases on assumptions rather than current evidence.

The fit test results should feed directly into two other tools introduced in this book. The first is the AI Efficiency Scorecard, which tracks performance metrics across all active AI use cases and needs to be updated whenever a fit decision changes. A use case that moves from expand to refine should immediately trigger a review of the metrics being tracked for that workflow, because the metrics that matter during a refine phase are different from the metrics that matter during an expand phase. The second is the 90 Day AI Process Optimization Plan, which sequences the improvement work that refine and pause decisions generate. When a fit test produces a refine decision, the specific improvements needed should be added to the optimization plan with owners, timelines, and success criteria, so the work gets done inside the operating rhythm rather than sitting on a list that nobody acts on.

The fit test cadence is what converts the AI Process Fit Test from a one-time diagnostic into a permanent operating discipline. A business that runs fit tests quarterly, documents the results, acts on the decisions, and adjusts its AI use cases based on current evidence rather than historical assumptions is building exactly the kind of AI operating discipline that produces measurable improvement over time. Fit scores will go up as workflows improve and standardization increases. Some will go down as business conditions change and new pressures appear. Both directions are useful. Both produce decisions the business can act on. And both are far more valuable than the alternative, which is running AI use cases indefinitely on the assumption that the initial approval is still valid because nobody has checked since.

Putting It Into Practice

AI should not touch every workflow simply because the tool is available or the team is eager to use it. That's the principle this chapter is built on, and it's the principle that separates businesses that use AI from businesses that benefit from it in ways they can measure and defend.

The five dimensions of the AI Process Fit Test give leadership a structured way to evaluate any workflow before expanding AI involvement, and a structured way to re-evaluate workflows that AI is already touching. Process stability asks whether the workflow is predictable enough for AI to produce consistent results. Output measurability asks whether the business has defined what good looks like.

Error tolerance asks how much risk the workflow can absorb if AI produces a flawed result. Handoff clarity asks whether responsibility at every stage is clearly assigned. Standardization level asks whether the workflow runs consistently enough for AI to improve it rather than accelerate its inconsistencies.

Those five dimensions produce a score. That score produces a decision: expand, refine, pause, or remove. Each decision has specific next actions, a clear communication approach, and a documentation requirement that keeps the AI operating record current and defensible. Walk-backs are not failures. They're evidence that leadership is governing AI with discipline. And a fit test cadence that runs quarterly, with named owners and results that feed into the AI Efficiency Scorecard and the 90 Day AI Process Optimization Plan, turns a one-time diagnostic into a permanent operating discipline that gets stronger every quarter it runs.

Now put it to work with three specific moves.

First, select your top three AI-supported workflows by activity level. Use the Workflow Reality Map findings from Chapter 4 as your evidence base. Schedule a two-hour scoring session for each workflow with a small group that includes someone who works inside the process, someone who manages it, and someone with a financial or operational perspective. Score all five dimensions, document the rationale, flag any dimension scoring a one, and record the total. Assign one of the four decisions to each workflow based on the scoring ranges in this chapter.

Second, for any workflow that receives a refine or pause decision, identify the specific dimensions that drove the score down and name the fix, the owner, and the timeline for each one. Don't produce a general improvement plan. Produce a specific list: this dimension scored a two because of this specific condition, this person is responsible for correcting it, and the correction will be complete by this date. That specificity is what turns a refine decision into actual workflow improvement rather than a documented intention that nobody acts on.

Third, assign a fit test owner to each active AI use case in the business and put the next quarterly re-score date on the leadership calendar now. The cadence only holds if it's scheduled. An owner without a deadline produces intentions.

An owner with a scheduled review date produces results. Before the next review arrives, ask the leadership team one question about each AI use case currently running: are we expanding this because the workflow is ready, or because the tool is already there? That question is the operating discipline this chapter is designed to install. It keeps AI placement decisions grounded in workflow evidence rather than tool availability, and it's the foundation on which every measurement chapter that follows is built.

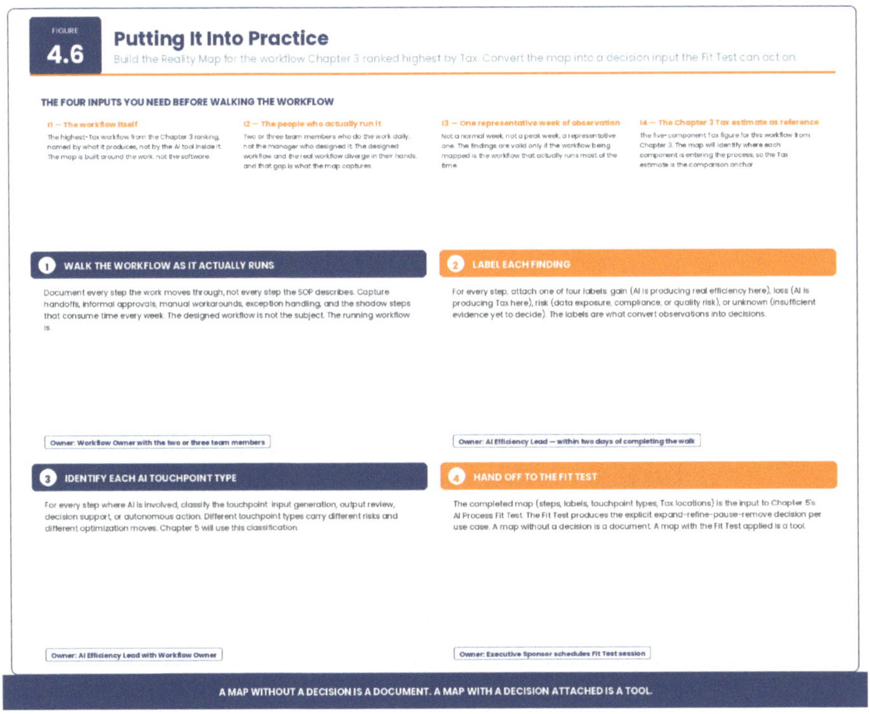

Chapter 6: The Four AI Performance Indicators

Measuring What Actually Moved

CHAPTER 6 OUTCOMES

At the end of this chapter, you will be able to:

1. Name the Four AI Performance Indicators: Cycle Time, Capacity, Error Rate, and Rework Cost — and explain why these four.
2. Capture or reconstruct a pre-AI baseline for each indicator, with documented confidence levels.
3. Select the right sample size for each indicator in an SMB context where statistical purity is rarely available.
4. Read the indicators together and interpret what to do when they disagree about whether AI is producing value.
5. Identify three common data quality failures in AI measurement and the hygiene practices that prevent them.

FOUR INDICATORS. THE SAME FOUR THAT VOLUMES I AND II MEASURED. NOW MOVING ON A DEFINED CADENCE.

The Problem This Chapter Solves: You Can't Manage What You Haven't Measured

At this point in the book, you've mapped your workflows, identified where the AI Efficiency Tax is running, and run the AI Process Fit Test on your highest-activity use cases. You have a clearer picture of where AI belongs and where it doesn't. That's real progress. But clarity about workflow conditions is still not the same thing as proof that AI is working. Proof requires numbers. Specific, documented, defensible numbers that show what actually changed inside the operation after AI was introduced.

Most leaders are still measuring the wrong things.

Usage rates, adoption percentages, output volume, and employee enthusiasm all tell you that AI is present. None of them tell you whether the business is faster, more accurate, more profitable, or less strained because of it. Those are different questions, and they require different measurements. The measurements that matter are the ones that show what actually moved inside the operation: cycle time, capacity, error rate, and rework (Safapour & Kermanshachi, 2019) . These four indicators are where AI performance either shows up or doesn't, and they're the only measurements that connect AI activity directly to the financial outcomes your CFO, board, and leadership team actually care about.

The measurement system in this Volume is built around those four indicators. The work that follows shows you how to establish a baseline, track changes over time, and connect each measurement to a financial result. The goal is straightforward: move AI reporting out of the activity column and into the performance column, where it belongs.

Without this chapter's work, everything else in this book produces insights without evidence. The Workflow Reality Map shows you what's happening. The AI Process Fit Test tells you what belongs. But this chapter is what tells you whether any of it is actually working, in numbers that hold up when someone in the room asks hard questions.

One operating distinction is worth making before the chapter goes further. Volume II Chapter 6 installed the Performance Reality Test and the activity-versus-outcome distinction as part of the AI Readiness and Performance Assessment. That work answers the readiness question: is the business measuring AI on the right indicators to make a defensible expansion decision. This chapter answers a different question. Once AI has been deployed and the readiness measurement work is in place, the leadership team needs four specific operating measurements running on a quarterly cadence to feed the AI Efficiency Scorecard in Chapter 7 and the AI ROI Formula in Chapter 8: cycle time, capacity, error rate, and rework cost. Volume II asks whether the business has the discipline to measure. Volume IV specifies which four measurements that discipline produces every quarter. The methodology Volume II established is the foundation. The four indicators this chapter develops are the operating outputs.

The Four AI Performance Indicators

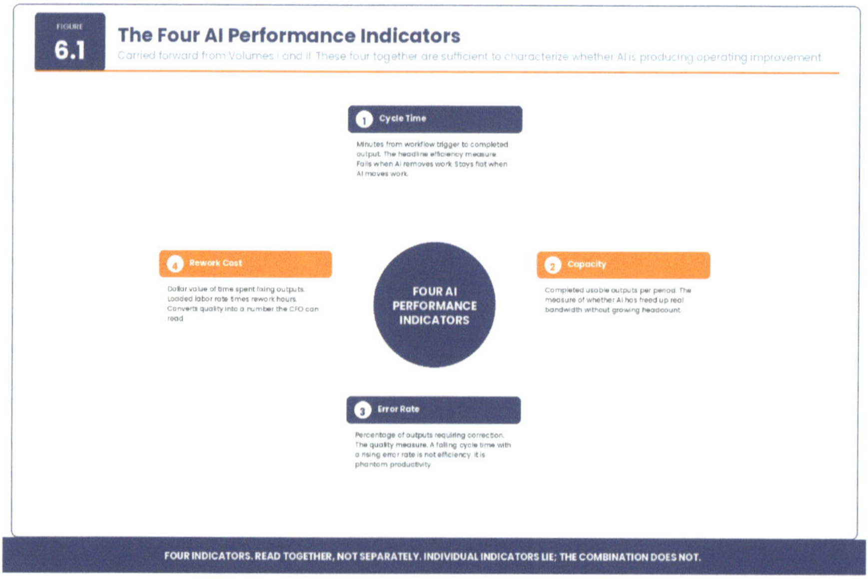

The set of four measurements this chapter develops, cycle time, capacity, error rate, and rework cost, is the operating instrument that every downstream framework in this book depends on. The AI Efficiency Scorecard scores these four indicators. The AI ROI Formula uses them to build the benefit side of the return calculation. The Executive AI Efficiency Brief reports on movement in these four. The 90 Day AI Process Optimization Plan tracks improvement against them. Without these four numbers, none of the downstream instruments produces a defensible result.

Each indicator answers a different question about AI performance. Cycle time asks whether AI made the work faster. Capacity asks whether AI freed productive hours that the business converted into more output. Error rate asks whether AI produced work that was usable the first time. Rework cost asks how much labor the business spent fixing what AI delivered. Four indicators. Four answers. Each one independent enough to stand alone, all four together producing a complete operating picture that no single measurement can produce on its own. The rest of this chapter walks through how to measure each one with the rigor that makes the number defensible.

Cycle Time Measurement. How to Track Whether AI Is Actually Making Things Faster

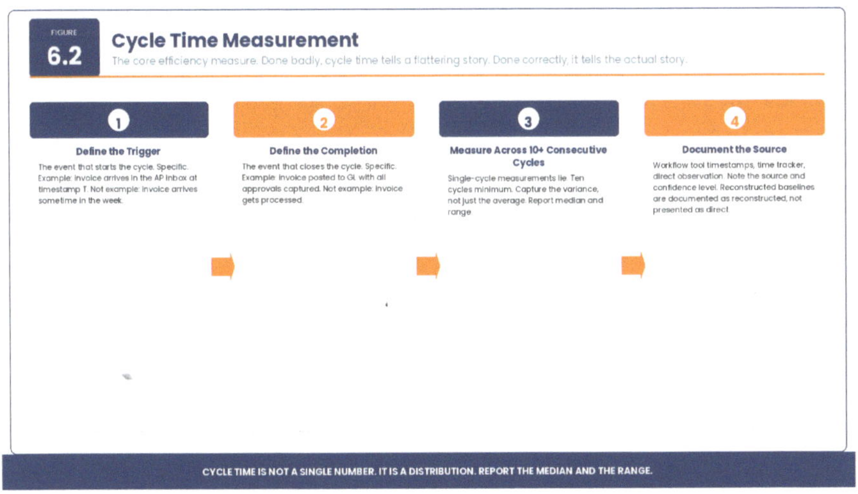

It's one of the clearest indicators of whether AI is improving speed or simply changing where the work happens. A workflow that completes faster after AI was introduced is showing a real efficiency gain. A workflow that completes in the same amount of time, or longer, is showing that AI shifted the work without removing it.

The first step in measuring cycle time is defining the start and end points precisely. This sounds obvious, but it's where most measurement attempts go wrong. The start point isn't when someone opens the AI tool. It's when the task enters the workflow, whether that's a client request arriving, a trigger event occurring, or a task being assigned. The end point isn't when AI produces its output. It's when the work is complete, approved, and delivered in a form that the business or the client can actually use. Everything in between those two points is the cycle, and that's what you're measuring.

Measuring only the AI step is one of the most common and most misleading mistakes in AI performance tracking. An AI tool might generate a draft in four minutes that used to take a team member ninety minutes to write. That looks like an enormous time saving. But if the draft then requires sixty minutes of review, twenty minutes of correction, and fifteen minutes of reformatting before it's usable, the total cycle time for that step has gone from ninety minutes to ninety-nine minutes. AI made the production faster and the workflow slower. Measuring the AI step in isolation would show a massive improvement. Measuring the full cycle would show a net loss.

To measure cycle time correctly, track the full workflow from start to finish, not just the step where AI is involved (Pierre et al., 2023) . For each cycle you measure, record when the task entered the workflow, when it exited as a completed output, and the total elapsed time between those two points. Do this for at least twenty consecutive cycles to get a reliable average. (Møller et al., 2017) Twenty cycles removes the distortion of outliers and gives you a number that reflects how the workflow actually performs, not how it is performed on one unusually fast or unusually slow day.

Compare the current average cycle time against a pre-AI baseline. If you captured a baseline before AI was introduced, use that number directly. If you didn't, reconstruct a baseline by asking the team to estimate how long the workflow took before AI was part of it, and document that estimate with the names of the people who provided it and the date the estimate was recorded. A reconstructed baseline is less precise than a captured one, but it's significantly more useful than no baseline at all. The direction of change matters more than the precision of the starting point in most SMB contexts.

When the Pre-AI Baseline Does Not Exist

Most businesses that adopted AI in the past two years did not capture a pre-AI baseline before AI was introduced. Volume II Chapter 6 walks through the reconstruction methodology in detail, including the three-question protocol with the team, the triangulation against indirect evidence (time-tracking records, project management logs, client invoices), and the documentation standard that produces a reconstructed baseline a CFO will accept as a working reference point. That methodology applies directly to the four indicators in this chapter, and there is no value in repeating it here. The Volume IV addition is the four-indicator specification: when reconstructing a baseline for cycle time, capacity, error rate, or rework, the team needs to produce a reconstructed estimate for each of the four indicators specifically, not a single composite estimate. Each indicator has its own reconstruction methodology, and each one feeds the Scorecard and ROI Formula independently. A reconstruction that produces one composite estimate but not the four indicator-level estimates is not yet usable for the work in Chapters 7 and 8.

Capacity and Error Rate Measurement. Tracking Throughput and Output Quality Together

Capacity tells you how much work the team can complete in a given period (Day et al., 2004) . Error rate tells you how much of that work is right the first time. These two measurements must always be evaluated together, because AI can increase throughput while simultaneously increasing mistakes, review burden, or inconsistency. Higher volume is only a win if the additional output is usable. If the additional output creates more correction work, client dissatisfaction, or internal rework, the volume increase has added cost rather than value.

Capacity measurement starts with a simple count. How many completed units of work does the team produce in a standard period, whether that's a week, a month, or a billing cycle? A completed unit is one that has passed review, met quality standards, and been delivered or approved for delivery. Work that's in progress, in review, or in a correction cycle doesn't count. Only finished, usable output counts. This distinction is important because AI tends to increase work in progress without always increasing finished output, and a measurement that counts in-progress work as completed capacity will overstate the efficiency gain significantly.

Measure capacity for the same period length before and after AI was introduced. If the team completed forty client reports per month before AI and now completes fifty-five per month, that's a thirty-eight percent capacity increase. That's a meaningful number. But before reporting it as a win, measure the error rate alongside it, because the capacity number without the error rate is only half the story.

Error rate is the percentage of completed outputs that required correction, revision, or rework before they were accepted as final (Patel et al., 2001) . Measure it by tracking how many outputs in a given period passed review on the first attempt versus how many required at least one round of correction. Divide the number requiring correction by the total number of outputs produced and express the result as a percentage. If the team produced fifty-five reports and twelve of them required correction before delivery, the error rate is approximately twenty-two percent. That's the number that needs to sit next to the capacity number in every performance report.

The combined picture of capacity and error rate tells you something neither number can tell you alone. A capacity increase with a stable or declining error rate is a genuine efficiency gain. AI is helping the team produce more usable work in the same amount of time. A capacity increase with a rising error rate is a warning signal. AI is helping the team produce more output, but a growing percentage of that output is wrong on the first pass, which means the review and correction cost is rising alongside the volume. At some point, the correction cost overtakes the throughput benefit, and the capacity increase becomes a net loss.

The threshold that matters most is first-pass accuracy. This is the percentage of AI-assisted outputs that are accepted without any correction. First-pass accuracy is the clearest single indicator of whether AI is reducing the total cost of producing work or simply shifting the correction burden to a different step in the workflow. A first-pass accuracy rate below seventy percent in most business workflows means the review and correction cost is likely consuming a significant portion (Major & Ngalamou, 2023) of whatever time AI saved in the production step. A rate above eighty-five percent generally indicates that AI is producing output clean enough to move through the workflow efficiently.

Set improvement targets that require both more capacity and better quality. This is the discipline that prevents the business from celebrating a volume increase that's quietly becoming more expensive to sustain. A reasonable target for a workflow that's been running AI for three to six months might be a fifteen percent capacity increase alongside a first-pass accuracy rate of eighty percent or better. Both conditions must be met for the target to count. Meeting only one of them means the workflow is trading one problem for another, not improving overall.

The financial connection here runs through labor cost and margin. Higher capacity without adding headcount means the existing labor investment is producing more output, which reduces cost per unit and improves margin. But that math only works when the error rate is stable or declining. If the error rate is rising, the correction labor cost is rising with it, and the per-unit cost may not be dropping at all despite the higher volume. Measuring both numbers together keeps the financial picture honest and prevents the business from reporting a margin improvement that the correction cost is quietly erasing.

Rework Measurement. Putting a Dollar Figure on the Cost of Doing It Twice

Rework is where AI savings disappear most reliably and most invisibly. Every time an AI output must be corrected, rewritten, verified, reformatted, escalated, or redone, the business spends labor hours that weren't included in the original promise of efficiency. Those hours are real. They're paid for through payroll. And in most businesses, they're never connected back to the AI tool that generated the need for them.

Rework has a narrow definition in most organizations and a much broader reality. The narrow definition is obvious correction: an AI output that was clearly wrong and had to be redone from scratch. That kind of rework is visible and usually gets noticed. The broader reality includes quiet cleanup work that employees treat as normal: adjusting the tone of an AI-generated email before sending it, adding the specific client context that AI didn't have, fixing the formatting that AI produced in a way that doesn't match the company's standards, removing the assumptions AI made that don't apply to this particular situation, and adding the judgment calls that AI can't make because they require knowledge of a relationship, a history, or a nuance that lives only in a person's head. That quiet cleanup is rework. It happens constantly. And because it feels like a normal part of using AI, nobody tracks it, nobody reports it, and nobody connects it to the AI tool that made it necessary.

Define rework broadly for measurement purposes. Any time a person spends modifying, correcting, supplementing, or rebuilding an AI output before it can be used, that time counts as rework. This includes both obvious corrections and quiet cleanup work. The only exception is intentional editing that was always part of the workflow design, where AI was explicitly brought in to produce a rough draft that a human would then refine as part of the standard process. In that case, the editing time is a designed step, not rework. Everything else that happens because the AI output wasn't good enough to use as delivered is rework, and it needs to be measured.

The measurement method is straightforward. Track how often rework occurs in each AI-supported workflow over a thirty-day period. For each rework instance, record the type of correction needed and the approximate time required to complete it. At the end of thirty days, total the rework instances and the total hours consumed. Multiply the total hours by the loaded labor rate of the person doing the rework. That calculation gives you the monthly rework cost for that workflow.

The loaded labor rate for this calculation should include salary, benefits, and overhead. For most SMB roles, the loaded rate runs between 1.25 and 1.4 times the base salary (Turner, 1919). Use the rate your CFO already applies to labor cost calculations. The goal is a number that uses the same methodology your financial team uses, so it's immediately credible when it shows up in a leadership conversation or a board report.

To illustrate the impact, consider a hypothetical example. A team of six analysts uses AI to produce weekly client performance summaries. Each analyst produces five summaries per week, for a total of thirty summaries per week across the team. On average, each summary requires twelve minutes of cleanup work before it's sent, adjusting tone, adding context, correcting one or two factual errors, and reformatting a section that AI consistently produces in the wrong structure. Twelve minutes per summary, thirty summaries per week, equals six hours of rework per week. At a loaded labor rate of sixty-five dollars per hour, that's three hundred and ninety dollars per week in rework cost. Annualized, that's just over twenty thousand dollars per year in labor cost that the business is spending to fix AI output in a single workflow. That number doesn't appear anywhere on the budget. It's absorbed into the analysts' normal working hours. But it's real, it's recoverable, and it shows up the moment someone measures it.

Rework is both an operational signal and a financial warning. Rising rework in a workflow that was previously stable usually means one of four things: the workflow inputs have changed and AI is no longer receiving the context it needs to produce accurate output; the prompting standards have degraded as team members drift from the approved approach; the use case has been expanded to a task type that AI handles less reliably than the original application; or the review standard has shifted and output that was previously accepted is now being corrected more aggressively. Each of those causes has a different fix. Measuring rework doesn't just tell you what it costs. It tells you where to look for the cause, which is what makes it one of the most operationally useful measurements in this entire chapter.

Compare the monthly rework cost against the savings AI was supposed to create. If AI was projected to save fifteen hours of production time per month at the same sixty-five dollar loaded rate, that's nine hundred and seventy-five dollars in projected monthly savings. If rework is consuming six hours per week, or roughly twenty-six hours per month, at the same rate, the rework cost is sixteen hundred and ninety dollars per month. The use case is running at a net loss of over seven hundred dollars per month, and it looks like a productivity win on every usage report the business produces. Measuring rework makes that reality visible. Ignoring it makes it permanent.

When the Metrics Disagree: How to Read Tradeoffs Before Declaring Success

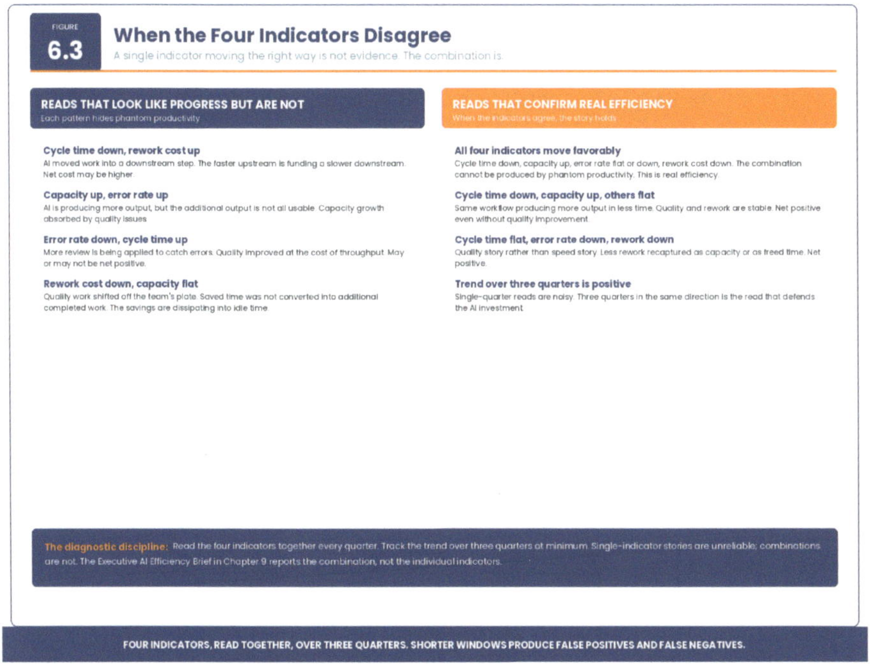

The most useful moment in AI measurement isn't when all four metrics improve together. It's when they don't. When cycle time drops but error rate rises, when capacity increases but rework grows, when first-pass accuracy improves but total cycle time stays flat, those disagreements are where the most important operating information lives. A leadership team that can read those tradeoffs accurately will make better AI decisions than one that declares success the moment any single number moves in the right direction.

Tradeoffs between metrics are normal. They're also informative. Each combination of improving and deteriorating metrics points to a specific operational pattern, and each pattern points to a specific type of fix.

Faster cycle time paired with higher error rate is the most common tradeoff pattern in AI-supported workflows. It means AI is accelerating the production step but producing output that fails quality checks more often. The workflow is moving faster on paper but generating more correction work downstream. The net result depends on how long the correction cycles are and how much they cost. If the correction cost is smaller than the time saved, the tradeoff is acceptable in the short term and worth improving over time. If the correction cost exceeds the time saved, the workflow is running a net loss that the faster cycle time is masking. The fix is almost always a combination of tighter prompting standards and an earlier quality checkpoint that catches errors before they enter the correction cycle.

Higher capacity paired with more rework is the second common pattern. AI is helping the team produce more output, but a growing portion of that output requires correction before it's usable. This pattern often appears when AI use cases are expanded to higher volume without first improving the standardization and quality conditions that support consistent output. The team is producing more, but the cost per usable unit is rising because the correction overhead is rising with the volume. The fix is to stabilize quality before expanding volume further, which usually means pausing the capacity expansion, tightening the output standards, and re-measuring first-pass accuracy before approving additional volume growth.

Lower review time paired with weaker output consistency is a subtler pattern and one of the hardest to catch without deliberate measurement. It appears when a business reduces review time by shortening the review process rather than by improving AI output quality. The review step gets faster because reviewers are checking less thoroughly, not because the AI is producing better work. Output consistency declines as a result, because errors that a thorough review would have caught are now passing through. This pattern is particularly dangerous in client-facing workflows because the quality problem reaches the client before anyone inside the business knows it exists. The fix is to re-establish review standards and measure first-pass accuracy separately from review time, so the business can see whether shorter review time reflects genuine quality improvement or simply reduced scrutiny.

The practical test for evaluating any AI use case is not whether one metric improved. It's whether the overall workflow became faster, cleaner, more reliable, and more financially defensible at the same time. That standard is harder to meet than a single metric improvement, and that's exactly why it's the right standard. AI should not be judged by whether it moved one number. It should be judged by whether it made the business better in a way that holds up across all four indicators without creating a hidden cost somewhere else.

When metrics disagree, the leadership response should follow a specific sequence. First, identify which metric is moving in the wrong direction and by how much. Second, determine whether the deterioration is large enough to offset the improvement in the other metrics, using the financial connection for each measurement to make the comparison in dollar terms rather than percentage terms. Third, identify the most likely operational cause of the deterioration based on the Workflow Reality Map findings and the fit test results. Fourth, make a decision about whether to adjust the workflow now or continue measuring for another thirty days to confirm the pattern before acting. And fifth, document the finding, the decision, and the rationale, so the AI operating record reflects how leadership responded to the tradeoff rather than just whether a metric improved.

Selecting the Right Sample Size for SMB Measurement

Statistical sampling theory was developed for large populations where pulling every instance is impractical. SMB measurement runs into a different problem. The population of instances for many AI use cases is small enough that the question is not how to sample but how to interpret the limited number of instances that exist. A practice firm processing eighty proposals per quarter does not need a statistically rigorous sampling methodology. It needs a defensible methodology for reading the eighty instances it has. The discipline for sample size in SMB measurement is different from the discipline for sample size in enterprise statistics, and applying enterprise methodology to SMB data usually produces conclusions weaker than the data could support.

The practical thresholds worth holding onto are these. Thirty instances of an AI-assisted output across a measurement period is the minimum for any meaningful trend reading (Zheng & Fan, 2024). With fewer than thirty instances, the variation between instances tends to swamp the trend signal, and conclusions drawn from the data are unreliable. The thirty-instance minimum is not statistically rigorous in the academic sense, but it is operationally adequate for SMB decisions where the alternative is no measurement at all. One hundred instances is the preferred sample for confidence in conclusions. With one hundred instances, the variation tends to settle into a readable pattern (Piscitelli et al., 2025), and the leadership team can reasonably attribute the pattern to the workflow rather than to chance. Above five hundred instances, additional sample size produces diminishing returns for SMB decisions, and the measurement effort is better spent on additional workflows or additional measurement dimensions.

For workflows that do not produce thirty instances per quarter, the measurement window extends. A workflow that produces ten instances per quarter requires three quarters to reach thirty. The trade-off is that the measurement lag is longer, and the leadership team cannot make confident decisions in the first quarter or two. The corrective response is not to lower the sample-size standard but to acknowledge the lag honestly, treat the early-period readings as directional rather than conclusive, and wait for the sample to mature before making expand or pause decisions. A low-volume workflow can still be optimized; the optimization just runs on a slower cycle than a high-volume workflow runs on.

The selection method matters as much as the sample size. The defensible approach is to measure every instance in the period rather than to sample randomly. Random sampling is appropriate for very large populations where measuring every instance is impractical (Henry, 1990). For SMB volumes (typically under five hundred instances per quarter for most use cases), measuring every instance is feasible and produces better data than random sampling would. The discipline is to capture cycle time, error rate, and rework cost for every instance that runs through the workflow during the measurement period, not for a sampled subset. The complete-population approach produces sharper conclusions and avoids the methodological questions that random sampling introduces.

Measuring AI Output Quality Beyond Error Rate

Error rate is the most measurable quality metric for AI output, and the most prominent in the standard measurement methodology. It is also the narrowest. Workflows where the AI is producing output for human review and use need quality measurements beyond the error rate, because outputs that contain no factual errors can still be unusable for other reasons. Three additional quality dimensions are worth measuring directly in any workflow where AI output is materially affecting business operations: consistency over time, defensibility under examination, and reviewer agreement rate.

Consistency over time measures whether the AI output for similar inputs produces similar quality across a measurement period. Output that scores well in week one but poorly in week six, or that drifts in tone or substance across the measurement window, is a leading indicator of model drift, vendor product changes, or workflow inputs that are themselves changing. The measurement protocol is straightforward: select ten representative inputs from week one, capture the AI outputs, then re-run those same inputs in week six and compare. Material differences between the two periods indicate inconsistency. The measurement does not need to be exhaustive; ten paired comparisons across a quarter is usually enough to identify drift if it exists.

Defensibility under examination measures whether the AI output would hold up if the business were asked to defend it. The defensibility question is most relevant for outputs that affect customer decisions, employment decisions, financial reporting, or regulatory submissions. The measurement protocol is to pull a sample of AI outputs and ask whether the team could defend each one if a regulator, an auditor, or a litigant examined it. Outputs that the team could not confidently defend should be reduced or eliminated from the workflow, regardless of their error rate or production speed. The defensibility question is a useful complement to the Volume III governance work and produces evidence that supports the per-use-case governance dossier.

Reviewer agreement rate measures whether two qualified reviewers, looking at the same AI output independently, reach the same assessment of whether it is acceptable. Low reviewer agreement rates indicate that the workflow standards are unclear, the AI output is in a borderline quality range, or both. The measurement protocol is to have two reviewers independently score a sample of AI outputs against the standard and compare their scores. High agreement (above eighty percent) suggests the workflow standards are clear and the AI output quality is consistent. Low agreement (below sixty percent) suggests the standards need clarification or the AI output is too variable for reliable use, and the corrective action depends on the underlying cause.

Measurement Hygiene: Three Common Data Quality Failures

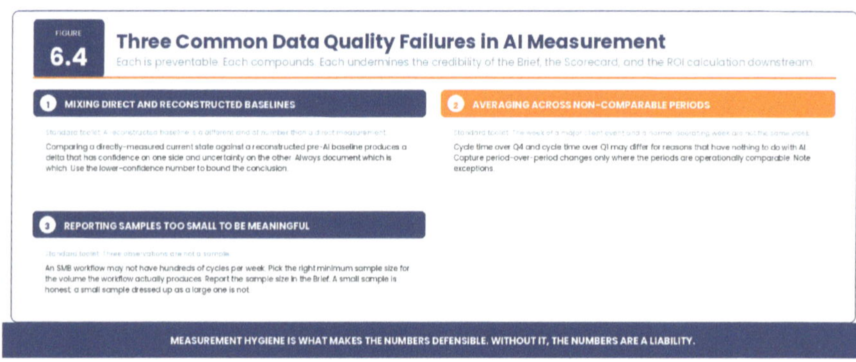

The measurement work in this chapter depends on the data the leadership team collects being clean enough to support conclusions. SMB measurement programs that have run for a quarter or two often produce data with three specific quality failures that erode the conclusions silently. The failures are not catastrophic, but they introduce enough noise that trend readings become unreliable and the team loses confidence in the measurement discipline. Recognizing the failures early prevents the larger problem of measurement that the team has stopped trusting.

Date stamp drift is the first failure. The data captures cycle time as the difference between the workflow start date and the workflow completion date, but the team records start and completion dates inconsistently. Some instances mark start at task creation, others mark start at first activity. Some instances mark completion at final approval, others mark completion at customer delivery. Without explicit definition, cycle time measurements vary by definition rather than by underlying performance, and trend readings become unreliable. The hygiene response is to document the start and completion definitions explicitly in the measurement methodology, train the team on the definitions, and audit a sample of instances each quarter to confirm the definitions are being applied consistently.

Owner drift is the second failure. The measurement record assigns each workflow to a named owner. Over time, the owner changes, the owner's reporting changes, or the owner role itself reorganizes. The historical data was attributed to one owner; the current data is attributed to another; trend readings across the change point are no longer comparable because the underlying owner has changed. The hygiene response is to document the owner attribution explicitly, note owner changes when they occur, and re-evaluate any trend reading that spans an ownership change with the change in mind. Ownership changes should not invalidate the data, but they need to be documented to keep the trend readings honest.

Definition drift is the third and most insidious failure. The workflow being measured at the start of the period is not exactly the workflow being measured at the end of the period. Steps were added or removed. The output specification changed. The input quality requirements shifted. The team did not recognize these changes as material at the time, so the measurement record continued to be captured as if the workflow was unchanged. Trend readings across the drift point are not comparing the same thing across periods; they are comparing different things that happen to share the same name. The hygiene response is to record material workflow changes in the measurement record at the time they occur, treat the periods before and after the change as separate measurement series, and avoid combining them into a single trend reading until the post-change period has produced enough data to read on its own.

Catching these three failures early is the difference between a measurement program that produces conclusions the leadership team trusts and one that produces numbers the team has learned to discount. The hygiene work takes about an hour per quarter for an established measurement program. The cost of not doing it is that the entire measurement program loses credibility, sometimes irreversibly, when the underlying data quality issues are eventually discovered. The hygiene discipline is one of the highest-impact practices in the Volume IV work.

The disciplines in this chapter convert measurement data into judgment. And judgment, grounded in documented evidence, is what the AI Efficiency Scorecard in the next chapter is designed to capture and report. The four measurements in this chapter are the inputs. The Scorecard is the output. The connection between them is the operating discipline that makes AI performance visible, defensible, and improvable over time.

Putting It Into Practice

This chapter walked through the four AI Performance Indicators and the methodology for capturing each one defensibly. The chapter's work is not finished when the methodology is read. It is finished when the four numbers exist on the top three AI-supported workflows, each one dated and named, ready to feed the Scorecard in the next chapter. Before turning the page, complete the four-step measurement pass below.

First, capture all four indicators on each of the three workflows the prior chapters identified. Cycle time uses the start-event and end-event methodology described above. Capacity counts completed, usable outputs per representative period. Error rate divides flawed outputs by total outputs against a documented quality definition. Rework converts correction labor into a dollar figure using the same loaded labor rate the Chapter 3 Tax calculation used. Every entry carries a date, a measurement person, a sample size, and a methodology note. Numbers without provenance survive no scrutiny when a CFO asks how they were built.

Second, reconstruct the baseline where one does not exist. Most SMBs introduced AI without capturing what the workflow cost before. The reconstruction methodology in this chapter is the defensible workaround: triangulate from team estimates, historical project records, and comparable workflow data. A reconstructed baseline produces a wider error band than a captured one, but it is substantially more useful than no baseline at all, and it is what makes the post-AI numbers comparable to a reference rather than floating without context.

Third, read the four indicators together rather than separately. A cycle time improvement paired with a rework increase is not a real improvement. A capacity gain paired with an error rate spike is borrowed throughput. The tradeoff section of this chapter explained how to recognize the patterns. The practical move at the leadership table is to ask, for each workflow, whether the four indicators are moving in the same direction. If they are not, the next operating decision starts with explaining why, not with declaring success.

Fourth, hand the four indicators to the Scorecard. The next chapter consolidates these numbers, the financial impact estimate from Chapter 3, and the Fit Test decision from Chapter 5 into a single recurring operating view the leadership team reviews each quarter. The Scorecard is what converts the indicators from a one-time measurement exercise into a permanent operating discipline that gets stronger every quarter it runs.

Four indicators. Four numbers. Each one captured with the methodology this chapter walked through, each one dated and named, each one ready to feed the AI Efficiency Scorecard in the next chapter. The measurement work is finished when the numbers are trusted enough that the leadership team is willing to make decisions on them and defend those decisions to the financial stakeholders who eventually ask. That trust comes from the rigor described above, and from one more thing the next chapter installs: the structured operating view that converts the four indicators into a single, defensible record of what AI is doing inside the business.

FIGURE 6.5

Putting It Into Practice

Capture the four indicators on each of the three priority workflows. Every number dated, named, and ready to feed the Scorecard.

THE FOUR INPUTS YOU NEED BEFORE THE MEASUREMENT PASS

I1 — The three workflows
The same top-three AI-supported workflows from Chapters 3, 4, and 5. Measurement focus stays with the workflows where the Tax was highest and the Fit Test decision was made.

I2 — The four indicators framework
Cycle time, capacity, error rate, rework. Defined the way this chapter defined them, not the way a vendor dashboard reports them. Definitions drive comparability.

I3 — A pre-AI baseline or a reconstructed one
Captured baseline if one exists. Reconstructed baseline from triangulated team estimates, historical records, and comparable workflows if it does not. A reconstructed baseline beats no baseline.

I4 — A two-to-four-week measurement window
Sample size adequate to absorb noise without spanning so much time that the workflow changes mid-measurement. The window length is documented alongside the sample size in the methodology note.

1 CAPTURE ALL FOUR INDICATORS

Cycle time from start-event to end-event. Capacity counts completed usable outputs per representative period. Error rate divides flawed outputs by total against a documented quality definition. Rework converts correction labor into a dollar figure at the loaded rate from Chapter 3. Every entry carries a date, a measurement person, a sample size, and a methodology note.

Owner: AI Efficiency Lead — captured across the measurement window

2 RECONSTRUCT BASELINE WHERE MISSING

Where no pre-AI baseline exists, triangulate from team estimates, historical project records, and comparable workflow data using the methodology in this chapter. Document the error band. A reconstructed baseline produces a wider band than a captured one, and it is substantially more useful than the alternative of no baseline at all.

Owner: Workflow Owner with AI Efficiency Lead

3 READ THE FOUR INDICATORS TOGETHER

A cycle time improvement paired with a rework increase is not a real improvement. A capacity gain paired with an error rate spike is borrowed throughput. For each workflow, ask whether the four indicators are moving in the same direction. If they are not, the next operating decision starts with explaining why.

Owner: AI Efficiency Lead — review with Workflow Owner

4 HAND THE NUMBERS TO THE SCORECARD

The next chapter consolidates the four indicators, the Chapter 3 financial impact, and the Fit Test decision into a single recurring operating view the leadership team reviews each quarter. The Scorecard is what converts the indicators from a one-time measurement into a permanent operating discipline that gets stronger every quarter it runs.

Owner: Executive Sponsor schedules Scorecard build

FOUR INDICATORS. FOUR NUMBERS. EACH DATED, EACH NAMED, EACH DEFENSIBLE.

Chapter 7: The AI Efficiency Scorecard™

A Single View of What AI Is Actually Doing for the Business

CHAPTER 7 OUTCOMES

At the end of this chapter, you will be able to:

1. Build an AI Efficiency Scorecard for each active use case using the five-category structure.
2. Use the Scorecard to make decisions, not just to report numbers — the difference between governance and reporting.
3. Recognize the five recurring score patterns in the Pattern Library and the operating action each pattern calls for.
4. Run the quarterly Scorecard review with the leadership team — a sixty- to ninety-minute session with documented outcomes.
5. Avoid the failure mode of building a Scorecard and then abandoning it — what abandonment looks like and how to prevent it.

THE SCORECARD IS HOW LEADERSHIP HOLDS AI ACCOUNTABLE TO THE SAME STANDARD AS EVERY OTHER OPERATING INVESTMENT.

The Problem This Chapter Solves: You Have Data but No Unified Picture

By the time a leadership team has worked through the tools in this book, something interesting happens. The business starts generating real AI performance data. Cycle time measurements from Chapter 6. Capacity and error rate numbers from the same chapter. Rework cost estimates from Chapter 3. Fit test scores from Chapter 5. Workflow findings from the Reality Map in Chapter 4. Individually, each of those data points is useful. Together, without a structure to hold them, they become noise.

That's the problem this chapter solves.

When AI performance data lives in separate spreadsheets, workflow notes, fit test documents, and informal estimates across different departments, leadership can't see the full picture. A CFO looking at one use case in isolation can't compare it against another. A COO trying to prioritize improvement work doesn't know which workflow deserves attention first. A CEO preparing for a board meeting can't quickly answer whether AI is generating net value or quietly consuming margin across the operation. The data exists. The unified view doesn't. And without a unified view, the business is still managing AI by feel, even when it has real numbers available.

The AI Efficiency Scorecard solves that by pulling scattered AI performance data into a single operating view. It shows every active AI use case in one place, scored consistently across the same five categories, with a decision recommendation attached to each one. It doesn't replace the detailed measurements from Chapter 6. It organizes them into a format that leadership can review, compare, and act on in a single sitting. The goal isn't another report. The goal is a decision tool that makes AI performance visible enough to manage, defensible enough to present, and simple enough to maintain inside the operating rhythm a lean leadership team already runs.

What Goes on the AI Efficiency Scorecard™ and Why

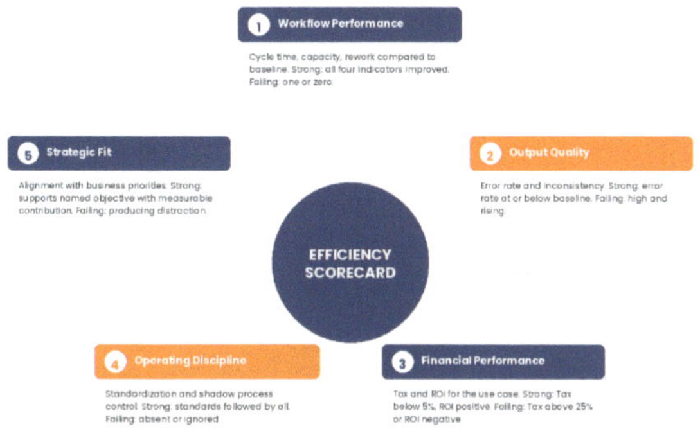

FIGURE 7.1 **The Five Categories of the AI Efficiency Scorecard**

Each scored on the same 3-2-1-0 scale as the Fit Test. The Scorecard is the operating-state version of the Fit Test.

SCORE EACH USE CASE QUARTERLY. THE PATTERN ACROSS THE FIVE CATEGORIES DRIVES THE OPERATING DECISION.

The AI Efficiency Scorecard tracks five categories for every active AI use case. Those five categories are cycle time change, capacity impact, error rate trend, rework frequency, and estimated financial impact. Every category connects directly to the operating questions that matter most to leaders who are accountable for results: did the work get faster, did the team gain capacity, did quality improve, did rework decline, and did the business see a financial benefit after all hidden costs were factored in? If you can answer those five questions for every AI use case in the business, you can manage AI as an operating discipline rather than a collection of experiments.

Cycle time change measures whether the workflow is completing faster, slower, or at the same speed compared to the documented baseline. Use the full-cycle measurement from Chapter 6, not the AI step in isolation. Score this category on a simple three-point scale: improving means cycle time has dropped by ten percent or more compared to baseline; flat means cycle time has changed by less than ten percent in either direction; deteriorating means cycle time has increased by ten percent or more. A simple three-point scale is deliberate. The goal isn't statistical precision. The goal is a clear directional signal that leadership can read in seconds and act on without needing to interpret a complex analytics model.

Capacity impact measures whether the team is handling more completed, usable output without adding headcount or extending hours. Use the capacity measurement from Chapter 6. Score this category the same way: improving means capacity has increased by ten percent or more per person; flat means capacity has changed by less than ten percent; deteriorating means capacity has declined or the team has added hours or headcount to maintain current output volume. The ten percent threshold is a practical operating standard for SMB workflows. (Agyei et al., 2026) It's meaningful enough to represent real improvement but realistic enough that a workflow can achieve it within a single quarter of focused optimization work.

Error rate trend measures whether output quality is getting better, staying the same, or getting worse over time. Use the first-pass accuracy rate from Chapter 6 as the primary indicator. Score it: improving means first-pass accuracy has increased by five percentage points or more since the last measurement period; flat means first-pass accuracy has changed by less than five points; deteriorating means first-pass accuracy has dropped by five percentage points or more. Error rate trend is the category that most often catches Phantom Productivity before it compounds, because a rising error rate is almost always the first signal that AI is producing more output without producing better output. (Deng, 2018)

Rework frequency measures how often AI output requires correction, modification, or rebuilding before it's usable, and what that correction work is costing in labor hours and dollars. Use the rework measurement from Chapter 6. Score it: improving means rework hours per week have declined by fifteen percent or more compared to baseline; flat means rework hours have changed by less than fifteen percent; deteriorating means rework hours have increased by fifteen percent or more. The fifteen percent threshold for rework is slightly wider than the ten percent threshold for cycle time and capacity because rework volume can fluctuate more week to week based on input quality and workflow conditions.

(Arashpour & Arashpour, 2015) A fifteen percent threshold filters out normal variation and surfaces only directionally meaningful changes.

Estimated financial impact is the category that converts the other four into business language. It's the net financial result of the use case after the AI tool cost, the review burden cost, and the rework cost are subtracted from the measurable savings in cycle time and capacity. Use the loaded labor rate calculation from Chapter 3 and the rework cost estimate from Chapter 6 to build this number. Score it: positive means the use case is generating net savings or margin improvement; neutral means the use case is roughly cost-neutral; negative means the use case is costing more than it's saving when all hidden costs are included. The estimated financial impact score is the one that CFOs and board members will look at first, and it's the one that most directly answers the question every leadership team eventually faces: is this AI investment paying off?

One critical principle applies to all five categories. The Scorecard should use simple, honest evidence rather than perfect data. A business that waits for perfect data before building its Scorecard will never build one. (Schrage et al., 2023) A business that uses honest estimates, clearly labeled as estimates, documented with the method used to produce them, will have a working operating tool that gets more precise every quarter as the measurement discipline matures. The strongest version of the Scorecard is the one that actually gets built, reviewed, and acted on. A theoretically perfect version that never gets completed is worth nothing.

To score each use case consistently, assign one person per use case as the measurement owner. That person is responsible for collecting the five category measurements, applying the scoring criteria, and bringing the scored Scorecard entry to each leadership review. Consistent ownership is what keeps the scoring honest. When different people score the same use case using different standards in different periods, the Scorecard becomes unreliable and comparisons across time periods lose their meaning. (Blasch et al., 2021) One owner, one standard, one documented method per use case.

Using the Scorecard to Make Decisions, Not Just to Report Numbers

A Scorecard that only reports performance is a reporting tool. A Scorecard that forces decisions is an operating tool. The difference between those two things is whether leadership uses the scores to make a specific choice about each use case every time the Scorecard is reviewed. If the review ends with a discussion but no decision, the Scorecard is adding a meeting to the leadership calendar without adding value to the operation. That's the outcome to avoid.

The AI Efficiency Scorecard is designed to support four specific decisions for each active AI use case: expand, refine, pause, or remove. These are the same four decisions introduced in the AI Process Fit Test from Chapter 5, and that's intentional. The fit test evaluates workflow conditions. The Scorecard evaluates performance results. Together, they give leadership two independent signals pointing toward the same decision, which makes that decision significantly more defensible than either signal alone could produce.

An expand decision is appropriate when three or more of the five Scorecard categories are scoring as improving, the estimated financial impact is positive, and no single category is deteriorating in a way that suggests a hidden cost is building. An expand decision means the business should invest more in this use case, whether that means processing higher volume through the same workflow, applying the same AI approach to additional similar workflows, or allocating more standardization and optimization work to deepen the efficiency gain that's already appearing. Document the specific expansion being authorized, the baseline scores at the time of the decision, and the next measurement checkpoint that will confirm the expansion is producing the expected improvement.

A refine decision is appropriate when the use case shows clear potential in some categories but has one or two categories scoring as flat or deteriorating. The use case isn't failing. It's underperforming relative to what the workflow conditions should support. A refine decision means the business will continue running the use case but will make specific, documented changes to the workflow, the prompting standards, the review structure, or the output quality criteria before the next Scorecard review. The refine decision must name what's being changed, who owns the change, and when the change will be complete. A refine decision without those three specifics is a postponed decision, not an operating choice.

A pause decision is appropriate when two or more categories are deteriorating, the estimated financial impact is neutral or negative, or when a single category is deteriorating severely enough to create risk for the business. Pause means AI stops touching that workflow until the underlying conditions are corrected. The pause decision should include a clear description of what needs to change, who is responsible for driving that change, and a specific date when the use case will be re-evaluated for re-entry. Pause decisions are not permanent. They're a disciplined response to evidence that the current workflow conditions can't support AI at the current level of involvement.

A remove decision is appropriate when the estimated financial impact is consistently negative across two or more Scorecard review cycles, when the fit test score is also low, and when no realistic corrective action would change the fundamental mismatch between the AI tool and the workflow. Remove means the use case is discontinued, the AI tool is taken out of that workflow, and the resources being spent on review, correction, and management overhead are redirected. Remove decisions require the most documentation because they're the most permanent. The documentation should include the Scorecard scores across all review cycles, the fit test scores, the specific reasons the use case is being discontinued, and the financial impact of continuing versus removing it.

The Scorecard should also connect back to the quarterly AI Process Fit Test results from Chapter 5. When a use case is scoring poorly on the Scorecard, the first question to ask is whether the poor performance is caused by weak process fit, poor measurement, rising AI Efficiency Tax, or flawed workflow design. A use case with a low Scorecard score and a low fit test score is telling leadership two things simultaneously: the workflow conditions aren't right and the performance results confirm it. That combination almost always points to a pause or remove decision. A use case with a low Scorecard score but a high fit test score is telling leadership something different: the workflow conditions are solid but the measurement or standardization work hasn't been done well enough yet. That combination almost always points to a refine decision.

The leadership discipline that makes the Scorecard valuable is non-negotiable. Every Scorecard review must end with a decision, an owner, and a due date for each use case that isn't already on an expand track. Without those three elements, the business is observing AI performance rather than managing it. Observation without action is how the AI Efficiency Gap stays open. (Iansiti & Lakhani, 2020) Management with documented decisions is how it closes.

Setting Up the Scorecard for the First Time When You Are Coming In Cold

A specific case deserves its own treatment. Some businesses arrive at this chapter without having completed the measurement work in Chapter 6, the fit test work in Chapter 5, or the workflow mapping in Chapter 4. They have AI use cases running. They have a leadership team that wants a Scorecard. They do not yet have the underlying evidence the Scorecard is designed to consolidate. The question is whether to wait until the prior chapters are complete, or to start the Scorecard with whatever evidence currently exists.

The right answer is to start now with what exists, mark every gap honestly, and let the Scorecard gaps drive the prior-chapter work. A first Scorecard that has three categories filled in for two use cases and not-yet-measured entries for everything else is more valuable than no Scorecard at all. It surfaces the gaps the business needs to close. It creates a visible artifact that leadership can refer to in the next operating meeting. And it sets the expectation that future Scorecards will be more complete, which produces accountability for completing the underlying measurement work.

The honest first Scorecard should record three things explicitly. First, which use cases are currently scored and what the scores are. Second, which use cases exist in the business but do not yet have enough data to score. Third, the named owner and target date for completing the missing measurements that will allow those use cases to be scored in the next cycle. That third item is what converts the gaps from observation to commitment. Without it, the gaps stay open quarter after quarter. With it, the second Scorecard is more complete than the first, and the third is more complete than the second.

This pattern is more common than most leadership teams expect, and admitting it openly is the right starting move. A business that builds its first Scorecard with three filled rows and eight not-yet-measured rows is not behind. It is starting. The discipline is what compounds. (Meyer, 2002) The first Scorecard is the proof that the discipline has started.

The Scorecard Pattern Library: Five Score Combinations Worth Naming

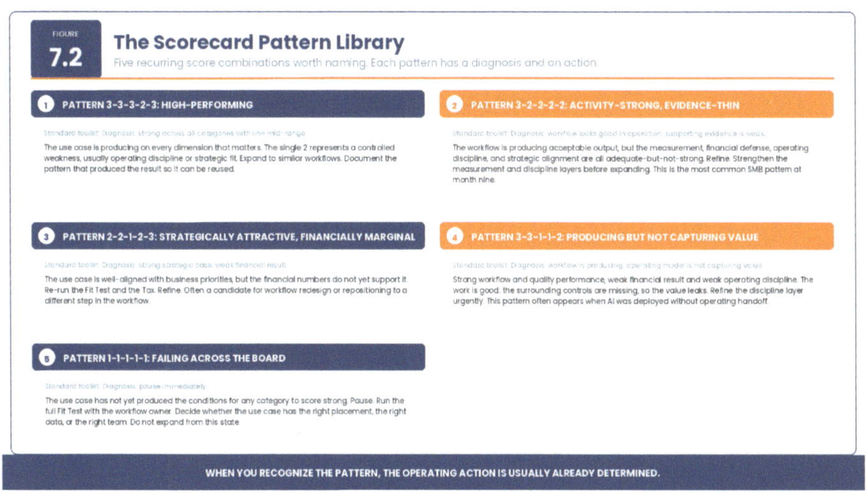

The five patterns described in this section are formalized in the Volume IV instrument that this section introduces: [PROPOSED MARK: The Scorecard Pattern Library]. The Library is the named collection of recurring score combinations the AI Efficiency Scorecard™ produces, paired with the operating decision each pattern requires. It is the bridge between raw category scores and the action those scores ask for, and it exists because a leadership team that has to interpret five separate category trends every quarter, for every use case, will eventually shortcut the interpretation in ways that erode the discipline. The Library removes that risk by naming what the team would otherwise have to derive from scratch every cycle.

The Library is built from the operating evidence Volumes I, II, and III did not produce. Volume I produced an inventory. Volume II produced a readiness decision. Volume III produced a governance dossier. None of those Volumes produced a recurring measurement record that could be interpreted as a pattern. The Scorecard Pattern Library only becomes possible once two or three quarters of Scorecard data exist for the active use cases, which is why it appears in this Volume rather than earlier in the canon. Every pattern named below assumes the underlying Scorecard discipline is running. The Library is not a substitute for that discipline. It is the interpretive layer that sits on top of it.

A leadership team that has run the AI Efficiency Scorecard™ for two or three quarters starts to recognize specific score patterns that recur across different use cases. Naming those patterns helps the team move faster through the review process and respond to each pattern with the operating action it actually requires. The five patterns below are not exhaustive. They are the most common combinations that surface in SMB AI operations after the measurement discipline has matured enough to produce reliable scores, and they cover the operating responses leadership will need ninety percent of the time. Other patterns will appear over longer measurement runs, and the Library is designed to be extended as the operating record grows.

The first pattern is improvement across all five categories with a positive financial impact. This is the strongest possible outcome. The workflow is producing more output, with higher quality, with less rework, in less time, at a net positive return. The correct response is rarely to celebrate. It is to extract the operational signal that produced this result and apply it to the next workflow. Strong scores are diagnostic. They tell the leadership team what conditions support real AI performance, which is exactly the information needed to replicate the pattern in workflows that are still underperforming.

The second pattern is improvement on cycle time and capacity with flat or deteriorating error rate. This is the most common false positive in AI optimization, and it is where Phantom Productivity hides on the Scorecard. The workflow is producing more output faster, but quality is declining. The financial impact line may still look positive because the cycle time improvements are dollar-quantifiable while the quality decline has not yet shown up as rework cost or client complaints. The correct response is to investigate why the error rate is moving in the wrong direction before approving any expansion, because expanding a workflow that is producing inconsistent output at higher volume multiplies the downstream cost faster than it multiplies the upstream benefit.

The third pattern is flat scores across all five categories. This is the pattern that most often gets misread as the use case being stable. It is more accurately read as the use case having reached the level of performance the current workflow conditions can support, with further improvement requiring changes to the workflow conditions rather than expecting the AI tool to extract more value on its own. A use case that scores flat for two consecutive quarters needs a fit test re-score and a Workflow Reality Map refresh, because the bottleneck has likely moved to a different part of the workflow than the one AI is currently optimizing.

The fourth pattern is improvement on rework and error rate with flat cycle time and capacity. This pattern often appears in workflows that recently completed standardization work. The standards are reducing inconsistency and lowering correction cost, but the cycle time and capacity gains have not yet appeared because the team is still adjusting to the new way of running the workflow. This is a setup pattern, not a problem pattern. Give it another quarter, then re-measure. If cycle time and capacity still have not moved, the standards are producing quality improvement without throughput improvement, which means the workflow design is the constraint and the next optimization cycle needs to focus on the workflow rather than on the AI step.

The fifth pattern is deterioration across two or more categories with neutral or negative financial impact. This is the clearest pause or remove signal the Scorecard produces. The workflow is consuming more cost while producing less value, and the trend is not moving in the right direction on its own. The correct response is a hard look at whether AI belongs in this workflow at all, using the AI Process Fit Test as the second signal. A use case showing this pattern alongside a low fit test score should be paused immediately. A use case showing this pattern with a high fit test score signals that the workflow has shifted in a way that broke the fit, and the workflow needs to be remapped before AI involvement is continued at the current level.

Naming these patterns has a practical effect on the leadership review. Instead of starting from the raw scores every quarter, the team can lead with the pattern that the scores match and move directly to the action it requires. That shortens the review, sharpens the decision, and keeps the Scorecard discussion focused on what to change rather than on what to describe.

The Pattern Library should be treated as a working document, not a fixed reference. Each quarterly review is an opportunity to confirm whether the existing patterns still match what the data is showing and to add new patterns when a use case produces a score combination the Library has not yet named. A business running the Scorecard for two years will accumulate three or four additional patterns beyond the initial five, each one tied to a specific operating cause and a specific response. That growth is what converts the Library from a list into an operating asset. Document each addition the same way the original five were documented: a name, a score combination, an operating cause, and a recommended decision. Over time, the Library becomes one of the strongest signals to a financial stakeholder reviewing the AI operations that the business is managing AI with measurement discipline rather than reacting to whichever metric happened to move that quarter.

One operational note about how the Library connects to the rest of the AI Operating System™. When a Scorecard review surfaces a pattern, the response is rarely confined to the Scorecard itself. A pattern that points to a workflow problem triggers a Workflow Reality Map refresh. A pattern that points to a fit problem triggers an AI Process Fit Test re-score. A pattern that points to a financial problem triggers an updated AI ROI Formula™ calculation. The

Library is the entry point that tells leadership which downstream tool to reach for next. That role makes it small, named, and disproportionately useful. It is the shortest path from observation to operating action that this Volume produces.

Reviewing the Scorecard With the Leadership Team

The AI Efficiency Scorecard becomes significantly more valuable when it's reviewed as part of the business's normal leadership rhythm. Not as a side project owned by one person. Not as an annual audit triggered by a budget conversation. As a standing agenda item in the same operating cadence the leadership team already runs, whether that's a monthly operating review, a quarterly leadership meeting, or a standing CFO briefing. When the Scorecard review lives inside the normal rhythm, it gets done consistently. When it lives outside the normal rhythm, it gets done when someone remembers to schedule it, which means it doesn't get done often enough to be useful.

The format for a Scorecard review is short and structured. Each AI use case owner presents four things: the current score across all five categories, what changed since the last review, which metric improved or deteriorated and by how much, and what decision they're recommending for the next period. The presentation for each use case should take no more than five minutes. A leadership team reviewing six active AI use cases should be able to complete a full Scorecard review in thirty to forty minutes. If the review is taking longer than that, the Scorecard has too many use cases on it or the presentations aren't focused enough on evidence and decisions.

The discussion that follows each use case presentation should focus on evidence rather than enthusiasm. The purpose of the review isn't to celebrate AI adoption or to defend AI investment. The purpose is to determine whether each use case is improving the workflow, consuming hidden costs, or requiring a change in process design. That distinction changes the tone of the conversation significantly. When the standard is evidence, the question isn't "is this use case working?" The question is "what does the data show, and what does the data require us to do next?" Those are different conversations, and the second one produces better decisions.

A useful review rule that keeps the Scorecard from becoming a static document: no use case should remain in the same status for more than two consecutive review cycles without showing documented improvement, being formally refined with a specific action plan, being paused, or being removed. If a use case has been sitting at the same flat or deteriorating score for six months with no meaningful change, that's not a stable situation. It's a situation where the business is tolerating underperformance because nobody has been willing to make the harder decision. The two-cycle rule forces the conversation. It requires leadership to either show that the use case is improving or make a more definitive choice about its future.

Consider a hypothetical scenario that shows what a productive Scorecard review looks like in practice. Return to the 75-person construction firm. Its CEO and leadership team have been running the AI Efficiency Scorecard for two quarters. At their most recent quarterly review, the Scorecard shows five active AI use cases. Two are scoring as improving across three or more categories and carry positive financial impact estimates. The team recommends expanding both. One is scoring flat across all five categories with a neutral financial impact. The use case owner recommends a refine decision, with a specific plan to rewrite the prompting standards and add a quality checkpoint earlier in the workflow. One is scoring deteriorating on error rate and rework, with a negative financial impact estimate. The use case owner recommends a pause, with a plan to remap the workflow before reintroducing AI. The fifth use case has been flat or deteriorating for two consecutive review cycles with no improvement despite a refine decision in the previous quarter. The team recommends a remove decision, documents the evidence behind it, and redirects the labor hours previously spent on review and correction toward the two expanding use cases. The entire review takes thirty-five minutes. Every use case has a decision, an owner, and a due date attached to it before the meeting ends. That's what a working Scorecard review looks like in an organization that's managing AI as an operating discipline rather than tracking it as an activity.

The CEO in that scenario isn't guessing about AI performance. He's reading evidence, making decisions, and holding his team accountable for results. His CFO can see the financial impact estimates. His COO can see where optimization effort should go next. His board can see a structured, documented record of how AI decisions are being made and what they're producing. That's what the Scorecard creates when it's built correctly and reviewed consistently.

How the Scorecard Connects to Annual Operating Planning

The Scorecard is most valuable when it feeds into the leadership team's annual operating planning rather than sitting as a standalone artifact reviewed only inside AI program meetings. Annual planning cycles in SMB businesses set the budget envelope, hiring targets, customer acquisition goals, and operating priorities for the next twelve months. The Scorecard contributes specific inputs to each of those decisions, and a leadership team that integrates the Scorecard into the annual planning rhythm extracts more value from it than one that runs the Scorecard separately. Three connection points matter most.

The first connection is to the budget envelope. The Scorecard contains every active AI use case, the financial impact of each one, and the ROI trajectory across the prior quarters. That information directly supports the AI line item in next year's budget. Use cases with strong and improving ROI deserve continued investment and possibly expansion. Use cases with weak or deteriorating ROI deserve scrutiny and possibly reduction. Use cases that score Phantom Productivity or that the fit test has tagged for pause should not be in next year's budget at the current level. The Scorecard converts the AI budget conversation from a vendor-driven negotiation into a performance-driven allocation, which is the conversation a CFO usually prefers to have.

The second connection is to hiring targets. AI use cases produce capacity effects that feed into hiring plans. A use case where capacity gains have been measured and confirmed reduces the next-year hiring requirement in the function the use case supports. A use case where capacity gains were assumed but not measured (a common pre-Volume IV pattern) often produces hiring shortfalls because the assumed capacity did not materialize. The Scorecard's capacity column is the most direct input to hiring planning, and the Scorecard data over multiple quarters provides the basis for confidence in the capacity assumption. Leadership teams that have run the Scorecard for two or more years often produce more accurate hiring plans than they did before the measurement discipline was in place.

The third connection is to customer acquisition and retention planning. AI use cases that improve customer-facing workflows (proposal turnaround, project status reporting, customer service response) feed into customer acquisition and retention metrics. A use case that improves proposal turnaround by twenty percent affects the close rate calculation in the sales plan. A use case that improves customer service response time affects the renewal rate calculation in the customer success plan. The Scorecard's connection to customer-facing metrics is more indirect than its connection to budget or hiring, but it is real, and leadership teams that integrate the Scorecard into customer planning often find that the AI investment is producing more customer-facing value than the operating-side measurement alone would have revealed.

The integration is not automatic. The Scorecard data feeds into annual planning when the leadership team builds the connection explicitly: scheduling the Scorecard review to land before the budget review, structuring the budget review to reference the Scorecard, and producing a documented record that connects Scorecard signals to specific planning decisions. The integration takes one to two planning cycles to mature. By the third cycle, the Scorecard is a standard input to annual planning rather than a separate AI artifact, and the value the measurement discipline produces compounds across the broader operating plan.

When the Scorecard Disagrees With the Team's Intuition

There are quarters when the Scorecard reports something the team is sure is wrong. The numbers say the workflow is underperforming, but the team is confident it is performing well. Or the numbers say the workflow is producing strong results, but the team has been frustrated with it for weeks. The disagreement between measurement and intuition is one of the most useful diagnostic moments in the Volume IV work, and the protocol for resolving it determines whether the Scorecard becomes a trusted tool or a contested one.

The first step is to take the disagreement seriously rather than dismissing either side. A measurement system that overrides team experience produces compliance but not trust. (Braithwaite & Makkai, 1994) A team intuition that overrides measurement produces narrative without evidence. Both are wrong as defaults. The right default is to treat the disagreement as a diagnostic signal that something needs investigation. The team is responding to information the measurement is not capturing, or the measurement is capturing information the team is not seeing, and the answer to which is which is the conversation that needs to happen.

The investigation runs through three specific questions. First, what is the team experiencing in the workflow that is not in the measurement record. Sometimes the team is observing quality issues that the error rate measurement does not capture (defensibility problems, customer feedback patterns, internal frustration about specific outputs). The investigation captures those observations in writing and decides whether to add them to the measurement record going forward. Second, what is the measurement record showing that the team has not noticed. Sometimes the team has been operating in the workflow without seeing aggregate patterns the data reveals (gradual error rate increases, drift in cycle time, capacity effects that take time to show up). The investigation walks the team through the data with the patterns highlighted and confirms whether the team agrees the patterns are real after seeing them. Third, given what both sides now see, what is the corrective action that addresses both signals.

The output of the investigation is one of three outcomes. The team's intuition was correct and the measurement is missing something important: update the measurement methodology. The measurement was correct and the team was not seeing the pattern: maintain the measurement and adjust the workflow accordingly. Both were correct about different things: address both signals through complementary corrective actions. The first outcome is the most common in the early quarters of measurement work, because the measurement methodology is still maturing. The second outcome becomes more common as the methodology stabilizes. The third outcome is the most useful when it occurs because it usually surfaces something the team genuinely did not know before.

The protocol prevents two failure modes that disagreement otherwise produces. It prevents the leadership team from overriding measurement based on team enthusiasm, which is the most common path to Phantom Productivity. And it prevents the leadership team from overriding team experience based on measurement, which is the most common path to a workforce that distrusts the measurement system and works around it. Either failure mode breaks the Volume IV work. The protocol holds both sides accountable and produces stronger outcomes than either side acting alone could produce.

The Scorecard at the Twelve-Month Mark

Four quarters of Scorecard data produces a different read than any single quarter does. The trends become visible. The seasonal patterns separate from the underlying performance. The team's response to corrective actions in earlier quarters shows up in current quarters as measurable change. The twelve-month Scorecard reading is the first moment in the Volume IV work where leadership can confidently distinguish what AI is producing for the business from what is noise, and it is the reading that should drive the most consequential operating decisions of the next year.

The four-quarter view answers questions that a single quarter cannot. Is the AI Efficiency Tax actually declining or just oscillating? The single quarter cannot tell. The four-quarter view can. Is the capacity gain real or seasonal? The single quarter cannot tell. The four-quarter view can. Did the standardization work from quarter two produce the rework reduction the team expected? The quarter-by-quarter data answers this directly. Is the ROI calculation stable across the year or driven by one strong quarter? The four-quarter view shows the variance.

The twelve-month Scorecard also surfaces use cases that need different attention going forward. Use cases with stable, strong ROI across four quarters can move to a lighter review cadence (semi-annual rather than quarterly) and free leadership attention for use cases that need more active management. Use cases with deteriorating trends across four quarters need formal review, possibly using the Process Fit Test from Chapter 5, and may need workflow redesign or removal. Use cases that have oscillated quarter to quarter with no clear trend need diagnostic work to identify what is causing the variance.

The twelve-month review is the natural moment to also revisit the Volume IV instruments themselves. Are the Scorecard categories still the right categories for this business, or have circumstances shifted in a way that warrants different metrics? Is the loaded labor rate methodology still defensible against current payroll structure? Is the AI Efficiency Tax calculation still capturing all the components that materially affect the number? The Volume IV work is built to be revisited and tuned, and the twelve-month review is the most useful moment to do that tuning. Leadership teams that treat the methodology as fixed often find the measurements drifting away from operational reality. Leadership teams that revisit and refine produce measurement that stays useful as the business changes.

One practical note on the review format: the Scorecard should be presented in a single document that fits on one or two pages. Not a slide deck with twenty-five slides. Not a spreadsheet with forty tabs. One document, one page per use case maximum, with the five category scores, the trend direction compared to the last review, the recommended decision, and the action items. The simpler the format, the more likely it gets reviewed consistently. A Scorecard that requires an hour of preparation to read is a Scorecard that gets skipped when the leadership team is busy. A Scorecard that can be read in five minutes per use case is a Scorecard that gets used every quarter without friction.

What Happens When a Scorecard Is Built and Then Abandoned

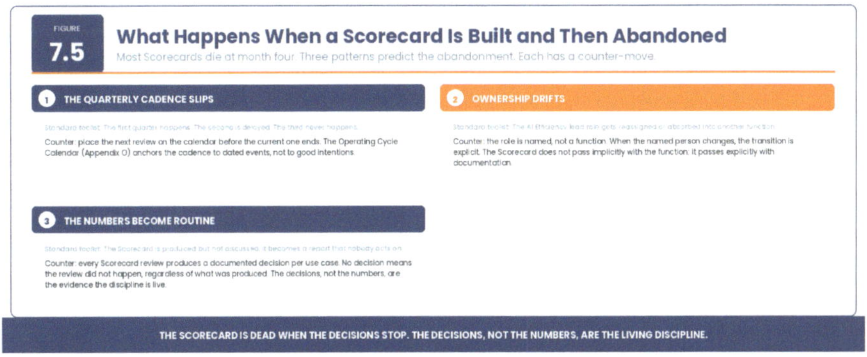

A 60-person professional services firm built its first AI Efficiency Scorecard with the energy of a new initiative. Three workflows scored, owners assigned, reviews on the calendar. By month four, the workflow owners were missing the review meetings. By month six, the Scorecard had not been updated. By month nine, when the managing partner asked to see the latest scores in preparation for a partner meeting, the document on the shared drive was the original quarter-one version. Three quarters of operating evidence had not been captured. Two use cases had quietly deteriorated. One had improved significantly and the firm had no record of why. The next conversation with the partners ran on impressions rather than evidence. The Scorecard was not the problem. The maintenance discipline was. A Scorecard without a recurring review cadence is a screenshot of a moment, not an operating instrument.

Putting It Into Practice

This chapter built the AI Efficiency Scorecard as the operating instrument that consolidates the prior chapters' work into a single leadership-facing view. The chapter's work is not finished when the document is built. It is finished when the Scorecard sits in front of the leadership team at the first scheduled review, with scores filled in, owners named, and a quarterly cadence on the calendar. Before turning the page, complete the four steps below.

First, populate the five Scorecard categories for every active AI use case. Cycle time change, capacity impact, error rate trend, rework frequency, and estimated financial impact, drawn from the Chapter 6 measurement record and the Chapter 3 Tax estimate. Document the rationale per entry, not only the score. Empty cells are findings, not gaps to gloss over. An incomplete Scorecard is more honest than a Scorecard with invented numbers, and it surfaces exactly where the next round of measurement work needs to focus.

Second, assign a named owner to each use case and schedule the first quarterly review on the leadership calendar before closing this action. The review is what converts the Scorecard from a document into an operating instrument. Without a scheduled review, the Scorecard becomes an artifact that gets referenced once and forgotten. With a scheduled review, the Scorecard becomes the recurring operating cadence that the next several chapters depend on.

Third, apply the pattern library to each use case. The five named patterns (clean win, hidden Tax, drift, deterioration, false negative) tell the leadership team what the score combination actually means in operating terms. Scores without pattern identification produce data. Scores with patterns produce decisions. The pattern is the bridge between the numbers and the action the leadership team takes next.

Fourth, hand the Scorecard to the Formula. The next chapter takes the Scorecard data and converts it into the financial number that a CFO, a board member, a lender, or an acquirer can hold up against any AI investment claim the business has ever made: the AI ROI Formula. The Scorecard is the operating view. The Formula is the financial view. Both are required, and the next chapter is where the second one gets built.

The Scorecard exists when the scores are filled in, the owners are named, the review cadence is on the calendar, and the document is sitting in front of the leadership team at the first scheduled review. Building it is the smaller half of the work. Running it is the larger half. The discipline that produces lasting AI value is not the initial build, it is the recurring review that turns numbers into decisions and decisions into measurable operating improvement.

Chapter 8: The AI ROI Formula™

Calculating a Number You Can Defend in Any Room

CHAPTER 8 OUTCOMES

At the end of this chapter, you will be able to:

1. Calculate the AI ROI Formula for any active use case using four benefit components and a complete cost denominator.
2. Distinguish the AI ROI Formula from the Volume III Governance Profit and Loss Formula and apply each to the right conversation.
3. Build a denominator that includes subscription, implementation, training, review burden, rework, management overhead, and governance effort.
4. Run sensitivity analysis on a ROI calculation before presenting it to a CFO, board, lender, or acquirer.
5. Defend the number under direct CFO challenge by tracing every input to a documented source.

A MODEST DEFENSIBLE ROI BEATS AN INFLATED ONE THAT FALLS APART UNDER FOLLOW-UP QUESTIONS.

The Problem This Chapter Solves: You Need a Number, Not a Story

There's a specific moment that arrives in almost every leadership team that has been using AI for more than a year. A CFO asks a direct question during a budget review. A board member wants more than a progress update before approving the next quarter's AI spend. A lender, an investor, or an acquirer asks what AI has actually done to operating performance before they make a decision. And the leader in the room, who has been championing AI for months, realizes they don't have a clean answer ready.

They have a story. They don't have a number.

The story sounds reasonable. The team is more productive. Workflows are faster. Employees are using the tools consistently. Adoption is high. The outputs look better than they did before. But none of that is a return on investment. None of it answers the question a CFO is actually asking, which is whether the money spent on AI produced measurable financial value the business can document and defend. Enthusiasm isn't a financial result. Activity isn't a return. (Sirota & Klein, 2013) And a story, no matter how well told, won't survive a serious financial review when someone in the room has a calculator and a skeptical disposition.

The AI ROI Formula is the practical mechanism that converts the measurement work from the previous chapters into a single, defensible return on investment figure. The formula draws directly from the cycle time measurements, capacity data, error rate trends, and rework cost estimates built in Chapters 6 and 7. It adds a complete cost denominator that includes every real expense the AI investment has generated, not just the subscription fee on the invoice. And it produces a number that can be presented to any financial stakeholder, explained in plain language, and defended under scrutiny because the methodology is transparent, the assumptions are visible, and the inputs are documented.

The goal of this chapter isn't to make AI look good. The goal is to calculate a number that's honest. A modest but defensible ROI number is worth more than an inflated one that falls apart the moment someone asks a follow-up question. (Milchanowski, 2026) The business that can walk into a boardroom with a real number, built on real evidence, with a clear explanation of how it was calculated, is in a fundamentally stronger position than the business that walks in with an impressive story and no math behind it.

By the end of this chapter, the phrase "AI is helping us" should be replaced with something far more useful: "AI returned this much, based on these inputs, after these costs, and here's the methodology."

The AI ROI Formula™. How It Works and What Goes Into It

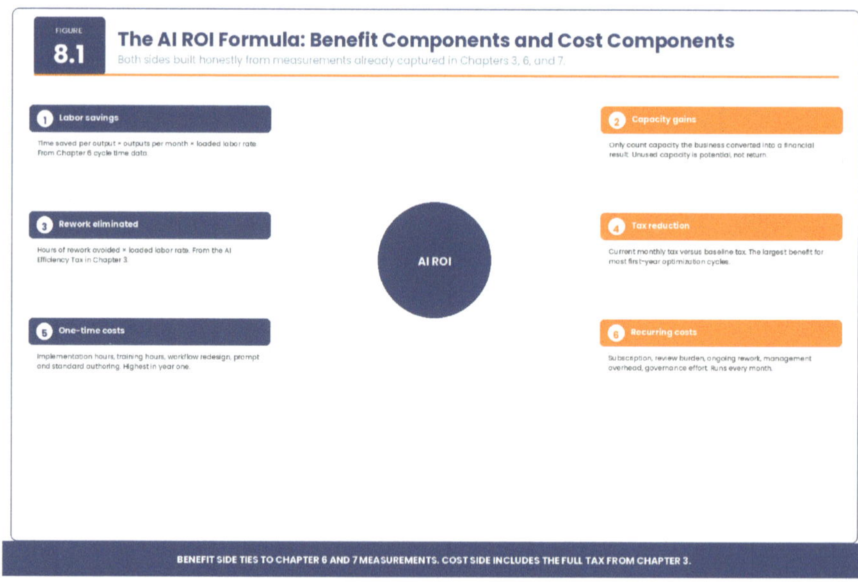

The AI ROI Formula converts operational improvement into financial value by comparing measurable AI benefit against total AI cost. In its simplest form, the formula is this: AI ROI equals total AI benefit minus total AI cost, divided by total AI cost, expressed as a percentage. That structure will be familiar to anyone who has calculated a standard return on investment before. What makes this formula specific to AI is the precision of what goes into each side of the equation, and the discipline required to build both sides honestly.

The benefit side of the formula has four components. Each one connects directly to the measurements built in Chapters 6 and 7, so the formula isn't a new calculation layered on top of the book's operating system. It's the natural result of the measurement work already done.

The first benefit component is labor savings from cycle time reduction. When a workflow completes faster because AI is involved, the time saved per cycle has a dollar value. Calculate it by multiplying the time saved per completed output by the number of outputs produced per month, then multiplying by the loaded labor rate of the people doing the work. If AI reduced the average cycle time for a specific workflow from five hours to three hours, saving two hours per output, and the team processes forty outputs per month at a loaded labor rate of seventy dollars per hour, the monthly labor saving from cycle time reduction is five thousand six hundred dollars. Annualized, that's sixty-seven thousand two hundred dollars. That's a real number with a real methodology behind it, and it belongs on the benefit side of the formula.

The second benefit component is capacity gains converted into financial value. This is the component that requires the most discipline, because capacity gains are only financially real when the business converts them into something measurable. Unused capacity is not a financial benefit. It's potential. (Ng et al., 1999) The distinction matters enormously when presenting this number to a CFO or board member. If AI freed up forty hours per month of team capacity and the business used those forty hours to serve additional clients, reduce overtime, avoid a new hire, or accelerate a revenue-generating project, the financial value of that capacity is real and calculable. If the forty hours were absorbed back into the workday without producing any additional output or avoiding any additional cost, the capacity gain has no financial value to report. Be honest about which situation applies. Report only the capacity gains the business actually converted into a financial result.

The third benefit component is error and rework cost eliminated. Use the rework cost calculation from Chapter 3 and Chapter 6 directly. If AI improvements to a workflow reduced monthly rework from twenty hours to eight hours, the twelve hours saved at the loaded labor rate of the people who were doing the rework is a real, recoverable financial benefit. Include both the direct rework time and any downstream costs that rework was generating, such as client correction cycles, delayed billing, or escalation management time. These numbers are already documented in the AI Efficiency Tax calculation from Chapter 3, which means they don't need to be rebuilt from scratch. They need to be pulled into the formula with a clear label.

The fourth benefit component is AI Efficiency Tax reduction. As the business standardizes workflows, improves prompting guidelines, redesigns review checkpoints, and removes shadow processes, the AI Efficiency Tax from Chapter 3 declines. That decline is a measurable financial benefit. Calculate it by comparing the current monthly tax estimate against the baseline tax estimate from before the optimization work began. The difference is the monthly tax reduction. Multiply by twelve for an annual figure. This component is often the largest single benefit in the formula for businesses that have been running AI without measurement discipline, because the tax was accumulating for months before anyone tracked it, and reducing it produces immediate, recurring margin recovery.

The total benefit figure is the sum of all four components, expressed as an annual number. Use annual figures consistently across both sides of the formula to make the ROI percentage meaningful and comparable across different use cases with different cost structures.

One important principle about capacity gains deserves emphasis before moving to the cost side of the formula. A capacity gain that the business didn't convert into a financial result should not appear on the benefit side. Leadership should resist the temptation to include theoretical capacity as a benefit because it makes the ROI number look stronger. A CFO who asks "did you actually use that capacity for anything, or did it just disappear into the workday?" deserves an honest answer. If the capacity was converted, report it with documentation of how. If it wasn't, leave it off the benefit side and note it as a future opportunity instead. That honesty is what makes the number defensible. Inflated benefit figures are the single most common reason AI ROI calculations fall apart under scrutiny.

What Counts as AI Cost. Building the Denominator Correctly

The denominator is where most AI ROI calculations become unreliable. This isn't because leaders are trying to mislead anyone. It's because the full cost of an AI use case is spread across multiple budget lines, absorbed into ordinary labor, and partially invisible (Bughin et al., 2017) in ways that make it genuinely difficult to see the complete picture without a deliberate effort to find it.

Most leaders count the subscription fee. That's the number on the invoice. It's easy to find and easy to report. But the subscription fee is almost never the largest cost in the denominator (Akpan, 2025) . The largest costs are the ones nobody labeled as AI costs when they were incurred: the implementation time, the training hours, the workflow redesigned work, the review burden, the rework cycles, the management oversight, the integration effort, and the AI Efficiency Tax that was running in the background before anyone started measuring it. Including only the subscription fee in the denominator produces an ROI number that looks impressive and means almost nothing, because it's comparing the benefit against a fraction of the real cost.

The complete cost denominator has two categories: one-time costs and recurring costs. Separating them matters because it affects how the ROI changes over time. One-time costs are front-loaded. They're highest in the first year and then they're gone. Recurring costs run every month. A use case with high one-time costs but low recurring costs will show a modest or negative ROI in year one and a strong ROI in years two and three. A use case with low one-time costs but high recurring costs, particularly if the recurring costs include a significant AI Efficiency Tax, may show positive ROI in year one and declining ROI over time as the hidden costs accumulate. Separating these two categories tells the business not just what the current ROI is, but whether it's improving or deteriorating as the use case matures.

One-time costs include implementation time, which is the total hours spent by internal team members setting up the AI tool, configuring it for the specific workflow, testing it, and troubleshooting it during the initial deployment. Value those hours at the loaded labor rate of the people involved. Include any external consulting or vendor implementation fees paid as part of the setup. Include the time spent mapping the workflow before AI was introduced, if that mapping work was done specifically to support the AI deployment. Include the cost of any workflow redesign work required to make AI fit into the process. And include the time spent creating initial prompting guidelines, quality standards, and review protocols. All of that work happened once, it had a real cost, and it belongs in the denominator.

Training hours are a one-time cost that almost never gets included in AI ROI calculations. When employees learn to use a new AI tool, that learning time has a dollar value. If six team members spent four hours each learning to use an AI writing tool, that's twenty-four hours of labor at the loaded rate of those employees. For a team of analysts at a loaded rate of sixty-five dollars per hour, that's fifteen hundred and sixty dollars in training cost. It's not a large number on its own. But across multiple AI use cases, training costs add up quickly, and leaving them out of the denominator systematically understates the true cost of the AI investment (Pandey et al., 2021) .

Recurring costs are where the denominator gets most seriously undercounted. The subscription fee is the obvious recurring cost. But the recurring cost category must also include the ongoing review burden, which is the time spent every month checking AI outputs before they can be used. It must include the ongoing rework cost, which is the time spent every month correcting AI outputs that weren't usable as delivered. It must include the management overhead cost, which is the time managers spend every month answering questions, resolving inconsistencies, and overseeing AI-supported workflows. It must include the governance documentation and compliance work required to maintain the AI operating record for that use case. And it must include any ongoing vendor management time spent on contract renewals, security reviews, or platform updates.

The AI Efficiency Tax from Chapter 3 is the most important recurring cost to include, and it's the one most consistently left out. The five components of the tax, review burden, rework cycles, output inconsistency cost, management overhead, and shadow process cost, are all recurring. They run every month. They're paid through payroll rather than through a vendor invoice. And because they're invisible in the budget, they're invisible in the denominator unless leadership makes a deliberate choice to include them. Including the full AI Efficiency Tax in the denominator is what separates a credible ROI calculation from an optimistic one. The CFO who reviews the calculation will be more impressed by a lower but complete ROI number than by a higher number that excludes the costs anyone with operating knowledge would know exist.

To build the complete denominator, use this specific sequence. First, pull the subscription or licensing cost from the budget for the full year. Second, estimate the one-time costs using the categories above and document each estimate with the person who provided it. Third, estimate the monthly recurring costs beyond the subscription fee using the AI Efficiency Tax calculation from Chapter 3, updated with the most current measurements from Chapter 6. Multiply the monthly recurring costs by twelve to get an annual figure. Fourth, add the one-time costs and the annual recurring costs together. That total is the denominator.

Then divide the total annual benefit by the total annual cost, subtract one, and multiply by one hundred to express the result as a percentage. That percentage is the AI ROI for that use case. If the benefit is one hundred and twenty thousand dollars and the complete cost is eighty thousand dollars, the AI ROI is fifty percent. If the benefit is sixty thousand dollars and the complete cost is seventy thousand dollars, the AI ROI is negative fourteen percent, which means the use case is currently costing more than it's returning. Both numbers are useful. The positive number tells leadership where to invest more. The negative number tells leadership where to stop investing or where to fix the cost structure before expanding further.

A practical note on precision: the denominator doesn't need to be perfect to be credible. It needs to be complete. A CFO reviewing an AI ROI calculation will accept reasonable estimates for items like management overhead and training hours, as long as those estimates are labeled as estimates, the methodology for producing them is documented, and the person presenting the calculation can explain how each number was derived. What a CFO won't accept is a calculation that excludes obvious costs because including them would make the ROI number look worse. Completeness is the credibility standard. Precision is secondary. (Bharadwaj et al., 2024)

How This Differs From the Governance Profit and Loss Formula in Volume III

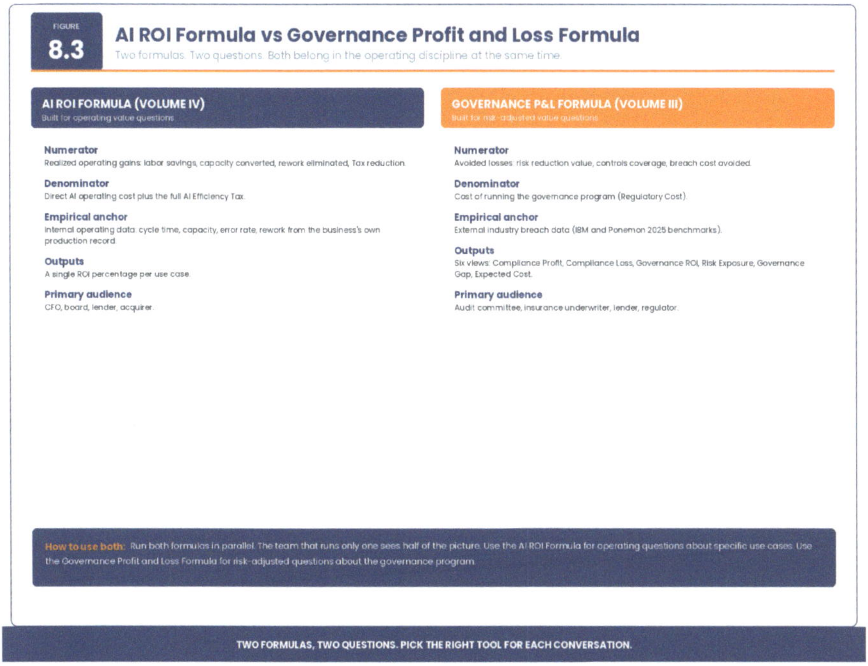

Readers who came through Volume III, The AI Risk and Governance Review, will recognize that this is not the first financial formula the Library introduces. Volume III installed the Governance Profit and Loss Formula, which uses the same financial vocabulary the CFO already speaks but applies it to a different question. Both formulas matter. Both should be running in the business at the same time. A leadership team that confuses one for the other will produce conclusions that the underlying math does not support. This section places them side by side, explains the structural differences, and gives the reader the operating reference for which formula belongs in which conversation.

The Governance Profit and Loss Formula from Volume III is actually a family of six formulas built on three risk-domain definitions. Regulatory Cost is the cost of building and maintaining the governance program, expressed in dollars.

Incident Cost is the cost the organization would bear from a governance failure on a given AI use case, including regulatory fines, remediation, legal exposure, reputational damage, and customer attrition. Governance Value is the risk-adjusted value the governance program produces by avoiding losses, preserving customer relationships, and supporting insurance recoveries when the controls operate as designed. From those three inputs, six working formulas follow.

Compliance Profit equals Governance Value minus Regulatory Cost, when Governance Value exceeds Regulatory Cost. This is the dollar return the governance program produced beyond what it cost to run, and it is the figure to take to the audit committee when defending the governance budget. Compliance Loss equals Incident Cost minus Governance Value, when Incident Cost exceeds Governance Value. This is the net loss the organization absorbed because the governance program did not produce enough value to cover the incident, and a persistent compliance loss on a specific use case is the signal to elevate that use case's tier or retire it. Governance ROI as a percentage equals Governance Value minus Regulatory Cost, divided by Regulatory Cost, times one hundred. This is the percentage return on the compliance investment. Risk Exposure as a percentage equals Incident Cost divided by Governance Value, times one hundred. This is the proportion of governance value at stake from a single incident on this use case, and a Risk Exposure above one hundred percent signals that the governance program does not, on its own, cover the downside scenario. Governance Gap measures the dollar cost or control-coverage percentage required to move from the company's current governance maturity to the applicable framework standard. Expected Cost of Ungoverned AI equals Incident Cost times Probability of Incident, which is the risk-weighted cost a CFO would book against a use case if the governance program were not in place.

The empirical anchor for the Volume III formula draws on the IBM and Ponemon Cost of a Data Breach 2025 figures, which give the CFO defensible benchmark values for Governance Value: roughly USD 192,000 saved per breach for governance technology, USD 147,000 for documented policies, USD 111,000 for board-level AI oversight, USD 114,000 for named security leadership, and

USD 223,000 for AI and machine learning insights in security operations. The Volume III formula draws on external industry breach data because the underlying question (what would a governance failure cost the company) cannot be answered from internal operating data alone. The breach has not happened yet. The benchmarks are how the CFO assigns a defensible number to the loss that the governance program is preventing.

The AI ROI Formula in this Volume is a single formula with four benefit components and a complete cost denominator. The benefit side captures labor savings from cycle time reduction, capacity gains the business actually converted into financial value, error and rework cost eliminated, and AI Efficiency Tax reduction. The cost denominator captures subscription cost, implementation time, training hours, ongoing review burden, ongoing rework cost, management overhead, and governance effort. The empirical anchor is the business's own measurement data from Chapter 6 (cycle time, capacity, error rate, rework) and Chapter 3 (the loaded labor rate and the AI Efficiency Tax). The Volume IV formula draws on internal operating data because the underlying question (what is this AI use case actually returning) can only be answered from the business's own production record.

The structural differences come down to five points. The numerators measure different things. The Governance Profit and Loss Formula's numerator measures avoided losses, expressed as risk reduction value. The AI ROI Formula's numerator measures realized operating gains. The denominators measure different things. The Governance Profit and Loss Formula's denominator is the cost of running the governance program. The AI ROI Formula's denominator is the direct AI operating cost plus the full AI Efficiency Tax. The empirical anchors are different. The Volume III formula draws on external industry breach cost data. The Volume IV formula draws on internal operating measurements specific to the business. The outputs differ. The Governance Profit and Loss Formula produces six different views of the same underlying data: Compliance Profit, Compliance Loss, Governance ROI, Risk Exposure, Governance Gap, and Expected Cost. The AI ROI Formula produces a single ROI percentage per use case.

The primary audiences overlap but are not identical. The Volume III formula is built principally for the audit committee, the insurance underwriter, the lender, and the regulator. The Volume IV formula is built principally for the CFO, the board, the lender, and the acquirer.

Both formulas are part of the AI Operating System. They run in parallel. They answer different questions about the same AI portfolio, using different data, producing different outputs, for partly different audiences. The leadership team that runs both formulas sees the full picture. The leadership team that runs only one sees half of it.

When to Use the AI ROI Formula and When to Use the Governance Profit and Loss Formula

The choice between the two formulas is not really a choice. The choice is which formula belongs in which conversation. The AI ROI Formula is the right tool for some questions. The Governance Profit and Loss Formula is the right tool for others. The leadership team that internalizes the distinction stops trying to make one formula do the work of both and starts pulling the right tool for each operating moment.

Use the AI ROI Formula when the question is whether a specific AI use case is producing operating value that exceeds its operating cost. This question arises in budget reviews when the CFO is deciding whether to continue funding a specific subscription or to reallocate the budget elsewhere. It arises in expand or pause decisions about individual use cases on the AI Efficiency Scorecard. It arises in the Executive AI Efficiency Brief when the business is presenting AI's contribution to financial performance for the period. The Governance Profit and Loss Formula will not answer this question because its numerator does not measure operating gains. Asking the Volume III formula about operating performance produces a number that has the wrong meaning for the conversation.

Use the Governance Profit and Loss Formula when the question is whether the governance program is producing risk-adjusted value beyond its cost. This question arises in audit committee reviews when the leadership team is defending the cost of the governance discipline. It arises in insurance renewal conversations when the carrier is asking what controls are in place and what they are saving the business. It arises in any situation where the leadership team needs to translate the work of governance into financial terms a board or external stakeholder will recognize. The AI ROI Formula will not answer this question because its numerator does not measure avoided losses. Asking the Volume IV formula about risk reduction produces a number that has the wrong meaning for the conversation.

Use Risk Exposure, from the Governance Profit and Loss Formula, when the question is what residual risk a single high-impact use case still carries after governance controls are applied. The AI ROI Formula does not produce a risk exposure measurement at all. Substituting AI ROI for Risk Exposure leaves the business exposed to incidents the AI ROI calculation will not predict, because a use case can produce strong operating ROI while still carrying material residual risk that the operating return does not address. The two questions are independent. Both need to be asked. Both need to be answered with the right formula.

Use the AI ROI Formula when the question is whether a use case is producing measurable margin improvement. Margin is an operating outcome, and the Volume IV formula is the one connected to the operating measurements (cycle time, capacity, rework cost, AI Efficiency Tax) that move margin. The Governance Profit and Loss Formula will not produce a margin signal because its inputs are not connected to those measurements. Margin improvement is a Volume IV question, and the AI ROI Formula is the right tool for it.

Use the Expected Cost of Ungoverned AI, from the Governance Profit and Loss Formula, when the question is what the company would lose if the governance program were not in place. This is the counterfactual question, and the AI ROI Formula does not address it. The Volume IV formula measures the actual operating result of the AI investment as the program is currently running. It does not measure what the alternative would be. The counterfactual is the territory of the Volume III formula.

Use both formulas together when the question is going to a board, a lender, an acquirer, or a regulator. The board wants to see the operating return (AI ROI) and the risk posture (Risk Exposure) in the same conversation. The lender wants both. The acquirer wants both. The regulator will be more interested in the Volume III outputs than in the AI ROI, but the combination of both produces a stronger position than either one alone, because the combination shows the leadership team understands both sides of the AI investment: what it is returning and what it is protecting the business from. A leadership team that walks into a high-stakes external conversation with both formulas in hand is in a substantially stronger position than one walking in with either one alone.

The combinations diagnostic is useful in its own right. A use case that scores well on both formulas is well-governed and productive, and it is the right candidate for expansion in the next operating cycle. A use case that scores well on the Governance Profit and Loss Formula but poorly on the AI ROI Formula is well-governed but is not earning its operating cost, and it is the right candidate for refinement or pause: the governance discipline is in place, but the operating economics are not yet working. A use case that scores well on the AI ROI Formula but poorly on the Governance Profit and Loss Formula is producing operating value but is under-governed and carrying residual exposure that the operating return alone does not justify, and it is the right candidate for governance reinforcement before further expansion: the business is taking a risk for a return that may not survive an incident. A use case that scores poorly on both is the right candidate for removal, and the combined evidence is stronger than either formula alone would provide.

The operating cadence runs both formulas on a quarterly cycle alongside the AI Efficiency Scorecard. The Governance Profit and Loss outputs feed into the per-use-case governance dossier from Volume III and into the AI Board Reporting Package. The AI ROI outputs feed into the AI Efficiency Scorecard from Chapter 7 and the Executive AI Efficiency Brief from Chapter 9. A leadership team running both formulas on the same cadence sees the full picture of the AI portfolio: the operating return and the risk exposure, side by side, calculated from compatible underlying data, and ready for any external stakeholder conversation the next quarter will bring.

The work in Volume III protects the business from what an AI failure can cost. The work in this Volume measures what AI is returning when it succeeds. Both numbers belong in the same operating conversation. Neither one alone is sufficient, and a leadership team that has both is operating the AI program at the discipline level the rest of the Library is built to install.

Presenting the AI ROI Formula Results to Financial Stakeholders

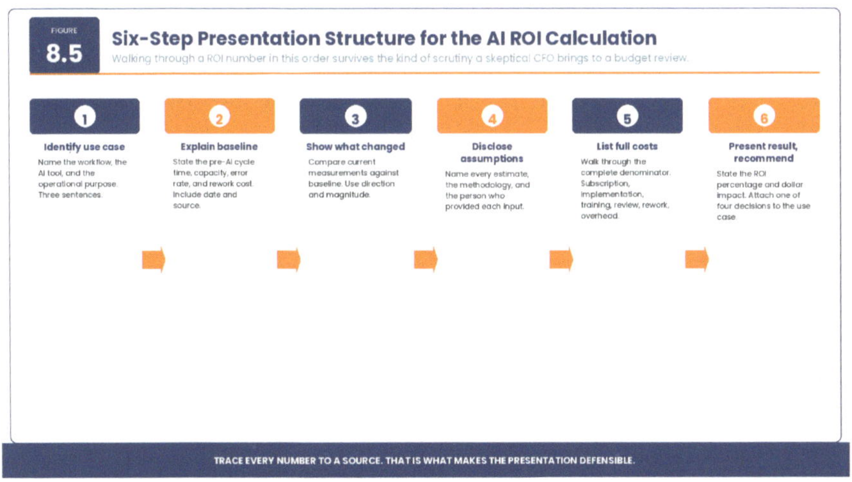

FIGURE 8.5 Six-Step Presentation Structure for the AI ROI Calculation
Walking through a ROI number in this order survives the kind of scrutiny a skeptical CFO brings to a budget review.

A strong ROI number only matters if leadership can explain it clearly under pressure. The number itself is the output of the formula. The presentation is what determines whether that number changes minds, earns trust, and produces the decisions the business needs. A well-calculated ROI presented poorly will generate skepticism. A well-presented ROI, even a modest one, will generate confidence. The difference is entirely in how the calculation is communicated.

The presentation structure that works best for financial stakeholders is simple and sequential. It has six steps. Use them in order every time.

Step one is to identify the use case. Name the specific workflow, the AI tool involved, and the team or department that uses it. Be specific. "Our AI-assisted contract review workflow in the legal operations team using the platform we deployed in Q2 of last year" is a better opening than "our AI tool for legal work." Specificity signals that the ROI is tied to a real, identifiable operating process, not a general claim about AI performance.

Step two is to explain the baseline. State what the workflow looked like before AI was involved. Cycle time, capacity, error rate, and rework cost at the pre-AI starting point. If the baseline was captured formally before deployment, say so and reference the documentation. If the baseline was reconstructed from team estimates after deployment, say that too. A reconstructed baseline with honest documentation is more credible than a claimed baseline with no documentation, even if the reconstructed number is less precise.

Step three is to show what changed. Present the current measurements for the same four indicators: cycle time, capacity, error rate, and rework cost. Show the direction and magnitude of change for each one. Where a metric improved, state by how much. Where a metric is flat or deteriorating, say that too. Presenting only the metrics that improved and omitting the ones that didn't will be noticed by any financially sophisticated stakeholder, and being caught omitting unfavorable data damages credibility far more than the unfavorable data itself would have.

Step four is to disclose the assumptions. Every ROI calculation rests on assumptions. The loaded labor rate used. The number of outputs per month. The percentage of capacity gains actually converted into financial value. The estimate methodology for management overhead. State each assumption clearly and briefly. This step is where most presenters get uncomfortable, because disclosing assumptions feels like exposing weakness. The opposite is true. A presenter who can state their assumptions clearly and explain why they chose them is demonstrating analytical discipline. A presenter who can't explain their assumptions is demonstrating that the number was produced without rigor.

Step five is to list the full cost categories. Show the denominator in detail. Subscription cost, implementation cost, training cost, review burden, rework cost, management overhead, governance effort, and any other recurring operational cost included in the calculation. Show the annual total. This is the step that most AI ROI presentations skip, and skipping it is the single most common reason those presentations fail under scrutiny. When a CFO asks "did you include the time your team spends reviewing AI outputs in that cost figure?" the answer should be "yes, and here's how we calculated it," not a pause followed by a recalculation.

Step six is to present the ROI result and state the recommended decision. Give the percentage. Give the annual net benefit in dollar terms. Then make a specific recommendation: this use case should be expanded, refined, paused, or removed, and here's why the ROI supports that recommendation. The presentation should end with a decision recommendation, not an open question. Financial stakeholders are not coming to the meeting to help leadership figure out what to do. They're coming to evaluate whether leadership has already figured it out. Walking in with a clear recommendation backed by a documented ROI calculation is exactly what that expectation requires.

Prepare for the questions that will come. The most common ones follow a predictable pattern. Were the labor savings actually captured, meaning did headcount go down, did overtime go down, or did the team handle more volume without adding people? Or did the time savings just disappear back into the workday? Did the capacity gain produce revenue, avoid a hire, or reduce overtime, or is it theoretical availability that nobody converted into anything? Did rework actually decline in a way that shows up in labor hours, or did employees just stop reporting rework because they stopped tracking it? Does the cost denominator include the management time spent overseeing AI outputs, or just the tool subscription?

Each of those questions has a direct answer if the measurement work from the previous chapters was done correctly. The cycle time measurements answer the first question. The capacity conversion documentation answers the second. The rework tracking logs answer the third. The complete denominator construction answers the fourth. The preparation for these questions isn't about anticipating attacks. It's about confirming that the ROI number is grounded in evidence that can be produced on request. If any of those questions can't be answered with documentation, that gap should be addressed before the presentation, not discovered during it.

The best presentation is calm and transparent. Overclaiming AI value is the fastest way to lose credibility with a financially sophisticated audience. (Kamraju, 2025) A CFO who catches an inflated assumption will discount everything else in the presentation, including the parts that are accurate. A modest, honest ROI number presented with complete cost disclosure and clear assumptions will earn more trust than an impressive number that can't survive a follow-up question. The goal of the presentation isn't applause for a strong AI story. The goal is to leave the room with the financial stakeholder's confidence and a clear decision about what happens next.

Using ROI Results to Make Portfolio Decisions Across All AI Use Cases

Once ROI is calculated across the active AI use cases in the business, something important becomes possible. Leadership can stop managing AI as a collection of separate experiments and start managing it as an investment portfolio. That shift changes how decisions get made, how resources get allocated, and how the business talks about AI performance at the leadership level.

A portfolio view means every active AI use case is evaluated against every other one using the same criteria. Return, risk, process fit, operational burden, and strategic importance. Those five dimensions give leadership a structured way to compare use cases that might look very different on the surface. A high-ROI use case in accounts payable and a high-ROI use case in client reporting are both strong performers, but they may carry very different risk profiles, very different process fit scores, and very different strategic implications for where the business is heading. The portfolio view holds all of that information in one place so that decisions about where to invest, where to pull back, and where to redesign are made in context rather than in isolation.

The portfolio management framework starts with a simple ranking. List every active AI use case in the business. Next to each one, record four numbers from the work already done in this book: the AI ROI percentage from the formula in this chapter, the AI Process Fit Test score from Chapter 5, the current AI Efficiency Scorecard decision from Chapter 7, and the estimated monthly AI Efficiency Tax from Chapter 3. Those four numbers together give a complete operating picture of each use case, and ranking them side by side immediately shows which ones are performing, which ones are struggling, and which ones are consuming resources without producing proportional value.

A use case with strong ROI and a high process fit score is the clearest candidate for expansion. The financial return is real, the workflow conditions support it, and the Scorecard is showing improvement across multiple categories. Expanding this use case means processing more volume through the same workflow, applying the same AI approach to adjacent workflows where the conditions are similar, or deepening the standardization and optimization work to extract more efficiency from what's already performing well. The expansion decision should be documented with the ROI number that supports it, the fit test score that confirms the workflow conditions, and the specific expansion being authorized.

A use case with weak ROI but genuine strategic importance requires a different kind of decision. Weak ROI doesn't automatically mean the use case should be removed. It means the use case isn't yet producing the financial return the business needs to justify its cost, and leadership has to decide whether the strategic importance justifies continued investment while the workflow is redesigned and the ROI improves. This is a judgment call that the ROI number informs but doesn't make on its own. The question to ask is whether a realistic redesign of the workflow, the prompting standards, or the review structure could produce a positive ROI within a defined timeframe. If the answer is yes, a refine decision with a specific improvement plan and a timeline for re-measurement is appropriate. If the answer is no, the strategic importance argument needs to be made explicitly and documented clearly, because spending money on a use case with negative ROI and no realistic path to improvement is a decision that deserves scrutiny every time the budget is reviewed.

A use case with negative ROI, a high AI Efficiency Tax, and a poor process fit score is the clearest candidate for removal. All three signals are pointing in the same direction. The workflow isn't right for AI, the financial performance confirms it, and the hidden cost is compounding every month the use case continues running. The remove decision here isn't a difficult judgment call. It's the logical conclusion of three independent data points all showing the same thing. The business should document the decision with all three numbers, communicate it clearly to the team using the framework from Chapter 5, and redirect the resources being consumed by the use case toward the ones that are performing.

Portfolio discipline is what prevents the business from funding AI because it's popular and redirects investment toward the places where AI is actually producing measurable value. Without portfolio discipline, AI budgets tend to grow based on enthusiasm rather than evidence. New use cases get approved because someone in the organization is excited about a new tool. Existing use cases continue receiving funding because stopping them would require acknowledging that they aren't working. And the overall AI investment keeps growing while the overall AI return stays flat or declines, because resources are being spread across too many use cases rather than concentrated in the ones that are demonstrably performing.

Sensitivity Analysis: Testing the ROI Number Before Presenting It

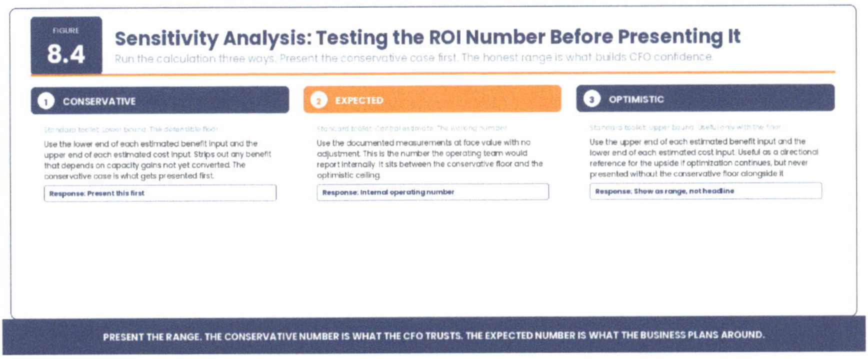

Any ROI calculation rests on assumptions, and the assumptions that go into the AI ROI Formula are not all equally consequential. Some assumptions move the result by a few percentage points if they shift. Others move the result by tens of percentage points. A leadership team that presents an ROI number without knowing which assumptions are doing the heaviest work in the calculation is in a weak position when a financial stakeholder asks what would happen if one of the assumptions turned out to be wrong. Sensitivity analysis identifies the load-bearing assumptions before the conversation reaches the room where the question would otherwise embarrass the team.

The mechanics of sensitivity analysis for the AI ROI Formula are straightforward. Take the calculated ROI percentage and identify the five inputs that contribute most heavily to it. For most SMB AI use cases, those five inputs are typically the loaded labor rate, the cycle time reduction estimate, the rework cost estimate, the AI subscription cost, and the ongoing review burden. For each of these inputs, the team adjusts the input value up by twenty percent and recalculates the ROI, then adjusts the value down by twenty percent and recalculates again. The output is a small table showing the ROI under three scenarios for each input: lower-bound, base case, and upper-bound. The table reveals which inputs the ROI calculation is most sensitive to.

The pattern that emerges is usually consistent across SMB AI use cases. The loaded labor rate is typically the most consequential input. A twenty percent change in the loaded labor rate often produces a thirty to fifty percent change in the calculated ROI, because the rate multiplies through every benefit component. The cycle time reduction estimate is typically the second most consequential. The rework cost estimate is typically third. The AI subscription cost is usually less consequential than expected, because it is a smaller component of the total cost denominator than the labor components are. The ongoing review burden is usually the fourth or fifth most consequential, depending on the workflow.

Knowing the sensitivity pattern changes how the team presents the ROI. Instead of presenting a point estimate (the workflow is producing 145 percent ROI), the team presents a range with the underlying drivers documented (the workflow is producing 120 to 165 percent ROI depending on labor rate assumptions, with the base case at 145 percent). The range is more defensible than the point estimate because it acknowledges the underlying uncertainty honestly. CFOs generally prefer ranges to point estimates for the same reason. The sensitivity table itself becomes a useful appendix to the Executive AI Efficiency Brief, because it shows the leadership team has examined the result for robustness before presenting it.

The three sensitivity tests every ROI calculation should pass before going to a financial stakeholder are these. First, what is the ROI if the loaded labor rate is twenty percent lower than the base case. If the result remains positive, the labor rate assumption is reasonably sound. Second, what is the ROI if the cycle time reduction is half of what the base case assumes. If the result remains positive or near-zero, the cycle time assumption is reasonably sound. Third, what is the ROI if the rework cost component is double the base case estimate. If the result remains positive, the rework cost assumption is reasonably sound. An ROI calculation that fails any of these three tests is producing a result that hinges critically on a single assumption, and the team should either tighten the underlying measurement or acknowledge the dependency in the presentation.

When the ROI Calculation Reveals That AI Is Not Yet Producing Net Value

Some ROI calculations come back negative. The use case is consuming more in cost than it is producing in benefit, and the math does not produce the result the leadership team was hoping to deliver. This is not a measurement failure. It is a measurement success. The calculation has surfaced what the team needed to know, and the next decisions are operating decisions rather than measurement decisions. The protocol for communicating a negative ROI honestly and acting on it constructively is one of the more useful disciplines in the Volume IV work.

The first principle is to lead with the number. Negative ROI in an AI use case is not a personal failure of the team or the leadership. It is a measured operating result, and the leadership team's job is to act on it. Burying the result in a narrative or surrounding it with caveats erodes the credibility of every other number in the AI program. Presenting it directly, in writing, with the methodology documented, produces a different conversation. The CFO who receives a negative ROI report with clean methodology and a proposed corrective action typically responds with engagement rather than alarm. The CFO who discovers a negative ROI buried in the narrative typically responds with broader skepticism toward the entire AI program.

The second principle is to separate the cause from the symptom. A negative ROI calculation tells you that costs exceeded benefits. It does not tell you why. The diagnostic work follows immediately. Was the cost denominator larger than expected because of unmeasured review burden or rework? Was the benefit numerator smaller than expected because the cycle time gains were not what the team thought? Was the use case the wrong fit for AI in the first place, which the Process Fit Test would have flagged earlier? The cause determines the corrective action. A use case where the cost denominator is the issue often has clear corrective actions in the standardization and process redesign work. A use case where the benefit numerator is the issue may need fit reassessment. A use case where the fit was always weak may need to be paused or removed.

The third principle is to commit to a specific decision within the next operating cycle. Negative ROI that produces no decision becomes accepted operating drag. The team learns that negative measurement does not produce action, and the measurement discipline weakens over time. The right pattern is to present the negative ROI, present the diagnostic conclusion, and present the corrective action with a timeline. The next Scorecard cycle then reports back on whether the corrective action moved the ROI in the right direction. Either it did and the use case is recovering, or it did not and a harder decision is needed. Both outcomes preserve the measurement discipline. The unacceptable outcome is to defer the decision indefinitely while the negative ROI continues to run.

Negative ROI is uncomfortable in the moment and useful in the longer arc. The Volume IV measurement discipline produces honest results. Honest results sometimes show that AI is not producing what was hoped. Leadership teams that handle those moments well build the credibility that makes the rest of the AI program defensible. Leadership teams that handle them poorly often find the measurement discipline itself coming under attack, which is the wrong outcome because the measurement was doing exactly what it was supposed to do.

The ROI Conversation With a Skeptical CFO

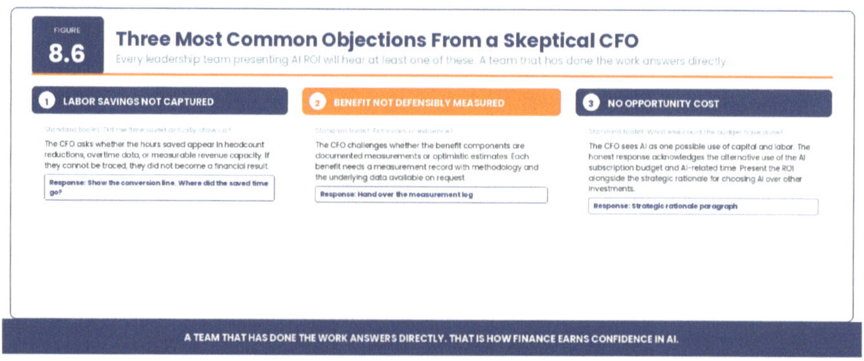

Some CFOs are predisposed to skepticism toward AI ROI calculations. The skepticism is usually justified by prior experience with AI claims that did not survive scrutiny, with vendor-supplied ROI figures that overstated the case, or with internal AI champions who presented optimistic numbers without the supporting methodology. The skeptical CFO conversation is not a hostile conversation. It is a conversation about whether the methodology behind the number is rigorous enough to act on, and the methodology is what determines whether the conversation goes well.

The most common CFO objection is that the cost denominator is incomplete. The skeptical CFO has seen AI ROI calculations that include the subscription cost but exclude the review burden, the rework cost, the management overhead, and the governance effort. (Singireddy, 2025) The objection is that the full cost of running AI in the business is substantially higher than the subscription, and any ROI that ignores the full cost overstates the return. The defensible response is to anticipate the objection and present the full cost denominator before being asked. A leadership team that volunteers the complete cost picture, including the components that other AI programs typically omit, establishes credibility that the skeptical CFO will recognize and respond to.

The second most common objection is that the benefit numerator is not connected to a defensible measurement. The skeptical CFO has seen ROI claims based on estimated time savings that were never verified, capacity gains that were assumed but not measured, and error rate reductions that the team reported without showing the underlying data. The objection is that the benefit side is built on optimistic estimates rather than measured outcomes. The defensible response is to ground every benefit component in a documented measurement, with the measurement methodology described and the underlying data available on request. A leadership team that can show the measurement record behind each benefit component answers the objection before it becomes a sticking point.

The third most common objection is that the ROI does not account for opportunity cost. The skeptical CFO sees the AI investment as one possible use of capital and labor, and asks what else those resources could have been doing. The defensible response is to acknowledge the question explicitly rather than dismissing it. The honest answer is that the alternative use of the AI subscription budget and the team's AI-related time is whatever next-best operating investment the business was deferring. The leadership team does not need to produce a formal opportunity cost calculation, but it does need to recognize that the ROI is being assessed against alternatives. Presenting the ROI alongside the strategic rationale for choosing AI over other investments closes this objection.

The pattern across these three objections is consistent. The skeptical CFO is asking whether the leadership team has done its work. A team that has done the work answers the questions directly, with documentation, and gains the CFO's confidence over time. A team that has not done the work attempts to deflect the questions and loses the CFO's confidence quickly. The investment in producing a defensible ROI calculation pays back disproportionately in the relationship with finance, because finance is the function most equipped to evaluate the methodology and most consequential when the AI program needs financial support.

The portfolio review should happen on the same quarterly cadence as the AI Efficiency Scorecard review from Chapter 7. In fact, the portfolio review is a natural extension of the Scorecard review, because the Scorecard already has the performance data organized by use case. Adding the ROI calculation and the fit test score to each Scorecard entry creates a complete portfolio view in a format that already fits the existing leadership rhythm. No new meeting required. No new reporting structure required. Just the addition of two columns to the Scorecard format and a portfolio-level conversation at the end of each quarterly review about where the overall AI investment is being allocated and whether that allocation reflects the actual performance evidence.

Putting It Into Practice

Before presenting any AI ROI number to a financial stakeholder, run it through one final check. Ask this question about every assumption in the calculation: would this number still hold up if the CFO challenged it directly? Not challenged it gently. Challenged it with the specific questions described earlier in this chapter, the ones about whether labor savings were actually captured, whether capacity gains were converted into real financial value, whether rework truly declined in a way that shows up in labor hours, and whether the cost denominator includes every real cost the use case has generated. If the answer to any of those challenges would require walking back a number, adjusting an assumption, or acknowledging that a cost was left out, fix it before the presentation, not during it.

That credibility test is the final operating discipline this chapter installs. It's not about being pessimistic about AI performance. It's about being honest enough that the number survives scrutiny. A number that survives scrutiny builds trust. Trust is what earns the approval to expand the use cases that are performing, the support to fix the ones that need redesign, and the credibility to make remove decisions without organizational resistance. All of that depends on the quality of the ROI calculation sitting at the center of the conversation.

Complete these four specific actions before moving into the next chapter.

First, calculate the AI ROI for your top three active AI use cases using the formula in this chapter. Build both sides of the equation completely: all four benefit components on the numerator and the full cost denominator including subscription, implementation, training, review burden, rework, management overhead, and governance effort. Write the calculation down with the methodology documented for each input. Date it. Name the person responsible for each number.

Second, create a one-page summary for each use case that follows the six-step presentation structure from this chapter. Identify the use case, explain the baseline, show what changed, disclose the assumptions, list the full cost categories, present the ROI result, and state the recommended decision. Keep each summary to one page. If it takes more than one page, the summary isn't focused enough yet.

Third, rank your three use cases by ROI percentage, fit test score, and recommended decision. Identify which one is the strongest candidate for expansion, which one needs redesign before further investment, and which one, if any, should be paused or removed based on the combined evidence. Write the ranking down with the rationale for each placement.

Fourth, run the credibility test on each calculation before sharing it with anyone outside the leadership team. For each assumption in the formula, ask whether it would hold up under a direct CFO challenge. Where it wouldn't, fix the assumption or document why the estimate is the best available evidence and what would be needed to make it more precise. The credibility test isn't a reason to delay the calculation. It's a reason to build it honestly from the start.

The work in this chapter gives the business something it almost certainly didn't have before: a documented, defensible AI return on investment number for each active use case, built on real measurements, with a complete cost denominator, and a clear decision recommendation attached to each one. That's the operating output that makes the next chapter's tool, the Executive AI Efficiency Brief, both possible and powerful. The Brief takes the ROI calculations, the Scorecard findings, and the portfolio decisions from this chapter and converts them into the communication format that any financial stakeholder, any board member, any lender, any investor, and any acquirer can read, understand, and trust. The number is ready. The next step is presenting it in the right format to the right audience at the right moment.

FIGURE 8.7

Putting It Into Practice

Calculate the AI ROI on the top three use cases. Run the credibility test before any number leaves the leadership team.

THE FOUR INPUTS YOU NEED BEFORE THE FORMULA RUNS

I1 — The top three use cases
The same three workflows the prior chapters worked on. The Reality Map, Fit Test score, four indicators, and Scorecard entry per use case all feed the Formula. The ROI calculation sits on top of the work the earlier chapters did, not next to it.

I2 — The four benefit components
Labor savings, capacity gains converted to financial result, rework cost eliminated, and Tax reduction. Only count capacity the business actually monetized. Unused capacity is potential, not return.

I3 — The complete cost denominator
Subscription, implementation, training, review burden, rework, management overhead, governance effort. Excluding any of these systematically overstates the return by a factor finance will identify in minutes.

I4 — The Chapter 3 loaded labor rate
The same loaded labor rate methodology used in the Tax calculation. Using one rate methodology in the Tax and a different one in the ROI denominator is the single fastest way to lose CFO confidence in the entire calculation.

1 CALCULATE AI ROI ON THREE USE CASES

Build both sides of the equation completely. Numerator: four benefit components. Denominator: full cost stack from subscription through governance. Write the calculation down with the methodology documented per input. Date it. Name the person responsible for each number. A calculation without provenance survives no scrutiny.

Owner: AI Efficiency Lead with CFO support

2 BUILD THE ONE-PAGE SUMMARY

For each use case, produce a one-page summary using the six-step presentation structure: identify the use case, explain baseline, show what changed, disclose assumptions, list full costs, present the ROI result, state the recommended decision. If it takes more than one page, the summary is not focused enough yet.

Owner: Workflow Owner with AI Efficiency Lead

3 RANK THE PORTFOLIO

Order the three use cases by ROI percentage, Fit Test score, and recommended decision. Identify the strongest expansion candidate, the one that needs redesign before further investment, and any that should be paused or removed. Write the ranking down with the rationale for each placement.

Owner: AI Efficiency Lead — reviewed with leadership

4 RUN THE CREDIBILITY TEST

For each assumption, ask whether it would hold up under a direct CFO challenge on the four pressure points (labor savings actually captured, capacity converted to financial value, rework truly declined in labor hours, every real cost in the denominator). Where it would not, fix the assumption before presentation, not during it.

Owner: AI Efficiency Lead before any external share

WOULD THIS NUMBER HOLD UP IF THE CFO CHALLENGED IT DIRECTLY? IF NOT, FIX IT BEFORE THE PRESENTATION.

Chapter 9: The Executive AI Efficiency Brief

Communicating AI Performance to the People Who Need to See It

CHAPTER 9 OUTCOMES

At the end of this chapter, you will be able to:

1. Populate the five sections of the Executive AI Efficiency Brief from work already completed in Chapters 4 through 8.
2. Tailor the same Brief for four different audiences (CFO, board, lender, acquirer) without changing the underlying data.
3. Assemble the six-document support packet that proves every number in the Brief is traceable to a source.
4. Run the three-cycle iteration pattern that turns a first imperfect Brief into a steady-state quarterly instrument.
5. Recognize the three categories of mid-quarter event that warrant an off-cycle Brief update without weakening the quarterly cadence.

ONE-PAGE BRIEF ON TOP. COMPLETE EVIDENCE BEHIND IT. THAT IS WHAT EARNS CONFIDENCE.

You've done the hard work. You've mapped the workflows, estimated the AI Efficiency Tax, run the AI Process Fit Test, built the AI Efficiency Scorecard, and calculated a defensible ROI number for each active use case. The measurement discipline is in place. The evidence exists. And now comes the moment that all of that work was building toward: someone outside the day-to-day operation needs to understand what AI has done for the business, and they need to understand it quickly, clearly, and without a forty-five-minute explanation of your methodology.

That's where most leaders hit a wall.

They have strong data. They have real numbers. They have a documented, traceable record of AI performance across multiple use cases. But when they sit down to present that information to a CFO, a board, a lender, or an acquirer, the communication breaks down. Either the presentation is too detailed, burying the most important findings under layers of methodology (Mohseni et al., 2021) and measurement explanation that the audience doesn't need. Or it's too thin, summarizing AI performance in vague language that doesn't hold up when someone asks a follow-up question. Neither version serves the business. Both versions undermine the credibility of the work that produced them.

The Executive AI Efficiency Brief solves this.

It's a structured, one-page communication tool that takes everything built across the previous eight chapters and converts it into a format any financial stakeholder can read in under five minutes, trust immediately, and use to make a decision. It doesn't replace the detailed evidence behind it. It sits on top of that evidence as the clean, polished, decision-ready surface layer that leadership needs for any high-stakes conversation about AI performance. The Brief is what you hand to the CFO before the budget review. It's what you present to the board before the quarterly update. It's what you send to the lender, the investor, or the acquirer before they ask for more detail. It's the document that makes the business look like it's managing AI with discipline, because it is.

The Problem This Chapter Solves: You Have the Numbers but No Way to Present Them Cleanly

Having strong AI performance data is only useful if you can communicate it in a way that the audience can absorb and act on. A CFO sitting in a budget review doesn't want a walk-through of your Workflow Reality Map. A board member preparing for a governance discussion doesn't need to see every row of the AI Efficiency Scorecard. A lender evaluating your operating performance doesn't want to read your rework calculation methodology. What each of those stakeholders wants is a clear, structured summary that answers four questions: what is AI doing in this business, what has it returned, what costs or risks remain, and what is leadership doing next?

Those four questions are exactly what the Executive AI Efficiency Brief is designed to answer.

The Brief isn't a dumbed-down version of the work from the previous chapters. It's the executive-facing output of that work. Think of it this way: the scorecard, the ROI formula, the fit test results, and the workflow findings are the evidence. The Brief is the argument built on top of that evidence. A strong argument doesn't require the audience to read all the evidence to understand the conclusion. It presents the conclusion clearly, supports it with the most important evidence, and makes the deeper evidence available for anyone who wants to examine it further. That's exactly how the Brief functions.

The most common failure mode in AI reporting isn't a lack of data. It's a lack of structure. (Weiner, 2020) Leaders who have done serious measurement work often present it in the format it was produced in, which is detailed, technical, and optimized for the person who built it rather than the person who needs to act on it. A CFO who receives a twelve-page AI performance report with measurement methodology appendices will either skim it, misunderstand it, or set it aside entirely. A CFO who receives a one-page Executive AI Efficiency Brief with a clear summary of performance, a defensible ROI number, and a specific recommendation will engage with it, ask sharp follow-up questions, and make a decision. The Brief creates the second outcome. That's why it exists.

What follows builds the Brief from the ground up, shows how to populate each section from the work already done in this book, explains how to tailor it for different financial audiences without changing the underlying data, and describes how to prepare the supporting documentation that gives the Brief its credibility when someone asks to see the numbers behind it.

What Goes Into the Executive AI Efficiency Brief. The Five Sections

FIGURE 9.1 **The Five Sections of the Executive AI Efficiency Brief**

Each section answers a question every financial stakeholder asks. Every number traces to a documented source.

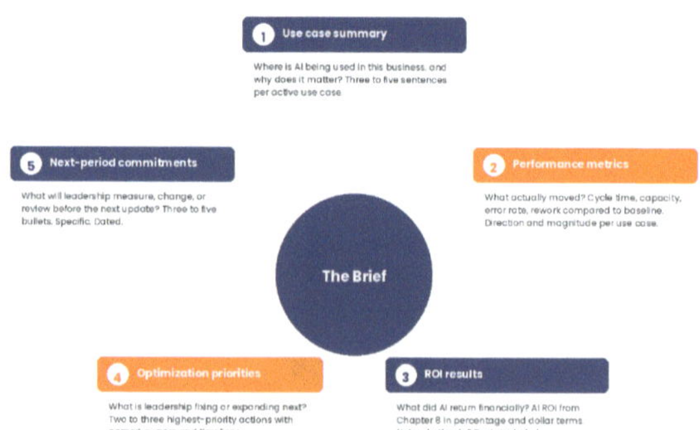

EVERY NUMBER IN THE BRIEF TRACES TO A DOCUMENTED SOURCE. THAT IS THE CREDIBILITY STANDARD.

The Executive AI Efficiency Brief has five sections. Each one answers a specific question that financial stakeholders ask. Each one draws directly from work already completed in the previous chapters. And each one is short by design, because the Brief's power comes from its clarity, not its length. A one-page document that every stakeholder reads and understands is worth more than a ten-page report that gets skimmed and set aside. (Bly, 2009)

The first section is the AI use case summary. This section answers the question: where is AI being used in this business, and why does it matter? It names each active AI use case, identifies the workflow it supports, and states the operational purpose it was deployed to serve. It should be three to five sentences for each use case, written in plain language that any financially literate reader can follow without operational context. Don't use tool names as the primary identifier. Use the workflow and the business function. "AI-assisted contract review in the legal operations team" is more meaningful to a CFO or board member than "our DocuSign AI integration." The use case summary is where the Brief establishes that AI is operating inside real, identifiable business processes, not floating as a general productivity tool across the organization.

The second section is the efficiency performance metrics. This section answers the question: what actually moved? It presents the current measurements for cycle time, capacity, error rate, and rework across each active use case, compared against the documented baseline from before AI was introduced. Use the measurements built in Chapter 6 and organized in the AI Efficiency Scorecard from Chapter 7. For each metric, show the direction of change, the magnitude of change, and whether the trend is improving, flat, or deteriorating. Keep this section factual and direct. One row per use case, four metrics per row, a clear comparison to baseline. This is not the place for interpretation or recommendation. It's the place for evidence. The interpretation comes in the sections that follow.

The third section is the ROI results. This section answers the question: what did AI return financially? It presents the AI ROI calculation from Chapter 8 for each active use case, including the total annual benefit, the total annual cost, and the resulting ROI percentage. It also shows the net financial impact in dollar terms, because a percentage without a dollar figure can obscure whether the ROI is meaningful at the scale of the business. (Mueller & Reardon, 1993) A fifty percent ROI on a ten-thousand-dollar investment is a five-thousand-dollar net return. A fifteen percent ROI on a two-hundred-thousand-dollar investment is a thirty-thousand-dollar net return. Both percentages tell a story. The dollar figures tell the business story. Include both. This section should also note whether the

ROI calculation includes the full AI Efficiency Tax or only the direct tool cost, because that distinction will come up in any serious financial review and it's better to disclose it proactively than to have it surface as a question that implies the calculation was incomplete.

The fourth section is the optimization priorities. This section answers the question: what is leadership fixing or expanding next? It identifies the two or three highest-priority actions coming out of the Scorecard review and the fit test results. For use cases receiving an expand decision, it states what the expansion entails and what outcome it's expected to produce. For use cases receiving a refine or pause decision, it names the specific problem being addressed, the action being taken, and the timeline for re-measurement. This section is where the Brief demonstrates that leadership is not just reporting AI performance but actively managing it. A financial stakeholder who sees a clear, specific list of optimization priorities with owners and timelines assigned is seeing evidence of operating discipline. That evidence is often as important as the ROI number itself, because it signals that the business knows what to do with the results it's producing.

The fifth section is the next-period commitments. This section answers the question: what will leadership measure, change, or review before the next update? It states the specific metrics that will be tracked over the next ninety days, the use cases that will be re-scored on the AI Process Fit Test, the workflow improvements that will be completed, and the date of the next Executive AI Efficiency Brief. This section is short, typically three to five bullet points, and it serves a specific purpose: it converts the Brief from a backward-looking performance report into a forward-looking accountability document. When leadership commits in writing to specific measurements, specific improvements, and a specific next review date, the Brief becomes a governance artifact, not just a communication tool. It creates accountability for the period between this Brief and the next one, and it gives any financial stakeholder a clear expectation of when they'll receive an update and what that update will cover.

These five sections together create a document that is short enough to read in one sitting, structured enough to answer every question a financial stakeholder is likely to ask, and traceable enough to defend when someone wants to see the evidence behind the summary. The Brief is one page. The evidence behind it may be fifty pages. The Brief is what gets read. The evidence is what gets trusted. Both are necessary. Only one goes into the room first.

One structural rule applies across all five sections without exception: every number in the Brief must trace directly back to a documented source in the underlying measurement record. Cycle time numbers come from the Chapter 6 measurement logs. ROI figures come from the Chapter 8 calculations. Fit test scores come from the Chapter 5 scoring sessions. Optimization priorities come from the Scorecard review decisions from Chapter 7. If a number in the Brief can't be traced to a specific documented source, it doesn't belong in the Brief. The Brief's credibility rests entirely on its traceability, and traceability requires that every number has a documented origin that leadership can produce on request.

Tailoring the Brief for Different Audiences: CFO, Board, Lender, and Acquirer

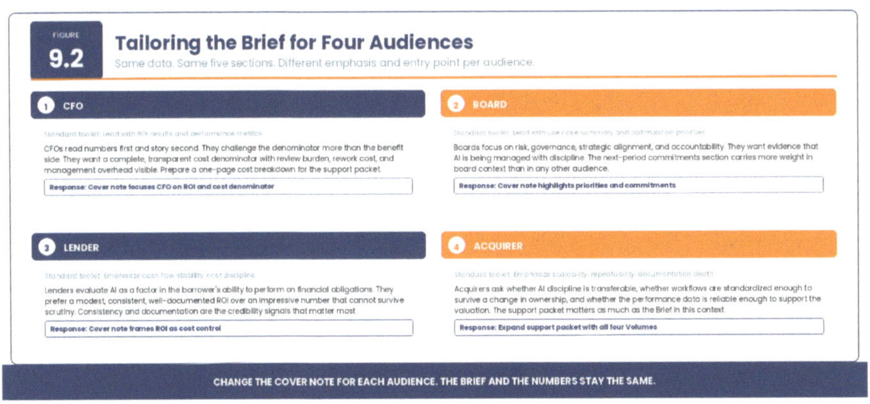

The same AI performance data supports very different conversations depending on who is sitting across the table. A CFO is asking different questions than a board member. A lender is looking for different signals than an acquirer. The Executive AI Efficiency Brief is flexible enough to serve all four audiences without changing the underlying data. What changes is the emphasis and the framing. The numbers stay the same. The lens through which those numbers are interpreted shifts based on what the audience cares about most.

This distinction is important enough to state directly. Tailoring the message for different audiences is smart communication. Changing the numbers for different audiences is a credibility problem (Rosenlund, 2025) that will surface the moment two different stakeholders compare notes. The Brief should present one measurement system, one set of results, and one set of decisions, viewed through four different stakeholder lenses. The data is constant. The emphasis is variable.

For a CFO audience, the Brief should lead with the ROI results section and the efficiency performance metrics. CFOs are financially trained and operationally literate. They want to see the numbers first and understand the story second. The most important questions a CFO will bring to the Brief are whether the labor savings were actually captured and show up in headcount or overtime data, whether the cost denominator is complete and includes the AI Efficiency Tax, whether the ROI calculation methodology is defensible, and whether the optimization priorities are financially grounded. When presenting the Brief to a CFO, be prepared to walk through the cost denominator in detail. CFOs are far more likely to challenge the denominator than the benefit side, because they know from experience that benefit figures tend to be optimistic and cost figures tend to be incomplete. (Hoitash et al., 2016) A CFO who sees a complete, transparent cost denominator that includes review burden, rework cost, and management overhead will trust the ROI number significantly more than a CFO who sees only the subscription fee in the cost column.

Prepare one supporting document specifically for CFO conversations: a one-page cost breakdown that shows every component of the denominator for each use case, with the methodology used to estimate each component and the person who provided each number. This document doesn't go into the room with the Brief. It stays in the support packet. But having it ready and being able to produce it immediately when the CFO asks "how did you calculate the cost side?" is the difference between a productive financial conversation and an uncomfortable one where leadership has to admit they'll need to follow up.

For a board audience, the Brief should lead with the use case summary and the optimization priorities, with the ROI results and performance metrics as supporting evidence. Boards are focused on risk, governance, strategic alignment, and accountability. (Drew et al., 2006) They want to know that AI is being managed with discipline, that leadership understands what's working and what isn't, and that there's a clear plan for improving performance over the next period. A board member reviewing the Brief will be most interested in whether AI use cases are being tested and evaluated rather than just adopted and forgotten, whether leadership is making clear decisions about which use cases to expand, refine, pause, or remove, and whether the next-period commitments are specific enough to create real accountability.

For board presentations, the next-period commitments section carries more weight than it does in any other audience context. A board that sees specific, dated commitments from leadership, with named owners and measurable outcomes, is seeing evidence of operating governance. That evidence is often what separates a board conversation that ends with confidence from one that ends with more questions than it started with. The optimization priorities section should also emphasize any use cases where a pause or remove decision has been made, because boards need to see that leadership is willing to make hard decisions based on evidence rather than protecting AI investments that aren't performing. A remove decision, presented with clear documentation and a straightforward rationale, demonstrates exactly the kind of disciplined AI governance that boards are increasingly expecting from leadership teams that have been using AI for more than a year.

For a lender audience, the Brief should emphasize cash flow stability, operating efficiency, and expense discipline. Lenders are not evaluating AI as a strategic asset. They're evaluating it as a factor in the borrower's ability to perform on financial obligations. The questions a lender brings to an AI performance conversation are whether AI is helping the business control costs, whether the AI investment is generating measurable operating improvement that supports cash flow, whether the AI Efficiency Tax is being managed or accumulating as a hidden liability, and whether the business is making disciplined decisions about AI spending rather than expanding it without accountability. When presenting the Brief to a lender, frame the ROI results in terms of cost control and operating stability. A lender is more interested in a modest, consistent, well-documented ROI than in an impressive ROI number that can't survive scrutiny. Consistency and documentation are the credibility signals that matter most in a lender context.

For an acquirer audience, the Brief should emphasize scalability, repeatability, documentation depth, and enterprise value contribution. An acquirer evaluating a business's AI operations is asking whether the AI discipline is transferable, whether the workflows are standardized enough to survive a change in ownership, whether the performance data is reliable enough to support the valuation, and whether the AI operating system adds to or subtracts from the business's enterprise value. The documentation behind the Brief matters as much as the Brief itself in an acquisition context, because acquirers conduct due diligence on the evidence, not just the summary. An acquirer who sees a clean Executive AI Efficiency Brief supported by a complete evidence packet, including Workflow Reality Maps, fit test scores, Scorecard records, and ROI calculations with full methodology disclosure, is seeing something rare: a business that has built AI into its operating discipline in a way that is documented, measurable, and defensible. That level of documentation doesn't just support the acquisition conversation. It supports the valuation.

The practical way to manage four different audience versions of the same Brief is to build one master Brief with all five sections fully populated, then create a brief cover note for each audience that frames the most relevant sections for that stakeholder's specific concerns. The cover note is one paragraph, written specifically for that audience, that tells them where to focus their attention in the Brief and why. The Brief itself doesn't change. The cover note changes the entry point. That approach takes less than thirty minutes per audience once the master Brief is built, and it ensures that every stakeholder is reading the same underlying data through the lens that's most relevant to their specific role and decision-making context.

Supporting the Brief With Evidence Without Overloading the Audience

The Executive AI Efficiency Brief is one page. It is never the only document in the room. Behind every Brief, there should be a support packet that contains the full evidence base the Brief summarizes. The support packet is not presented upfront. It's available on request. The distinction between those two things is the operating principle that keeps the Brief clean while keeping the overall communication credible.

Think of it as a two-layer communication structure. The Brief is the top layer. It's what every stakeholder reads. It's what drives the conversation. It's what gets shared, referenced, and acted on. The support packet is the second layer. It's what proves the Brief is traceable when someone asks where a number came from. The

Brief earns the initial trust. The support packet sustains that trust under scrutiny. Both layers are necessary. Only one goes into the room first.

The support packet should contain six specific documents. The first is the AI Efficiency Scorecard from Chapter 7, showing the current scores across all five categories for every active use case, the trend direction compared to the previous review period, and the decision recommendation attached to each use case. The second is the AI ROI calculation worksheets from Chapter 8, one per use case, showing both sides of the formula with full methodology documentation for every input. The third is the baseline measurement records from Chapter 6, showing the pre-AI and current measurements for cycle time, capacity, error rate, and rework for each workflow where AI is active. The fourth is the AI Process Fit Test results from Chapter 5, showing the scores across all five dimensions for each use case and the rationale for the scoring decisions. The fifth is the Workflow Reality Map findings from Chapter 4 for any use case where a refine, pause, or remove decision has been made, because those decisions require the most explanation and the map findings are the most direct evidence of why the decision was correct. The sixth is a brief decision log that documents every major AI decision made in the current period: use cases expanded, use cases refined, use cases paused or removed, workflow changes implemented, and standardization improvements completed. The decision log shows that leadership is actively managing AI rather than passively reporting on it, and it creates a clear record of the governance discipline that financial stakeholders are increasingly expecting to see.

The practical rule for using the support packet is simple. Present the Brief first. Answer questions from the Brief. Only produce a support document when the audience asks a specific question that requires it. A CFO who asks "how did you calculate the rework cost in the ROI formula?" gets the ROI calculation worksheet. A board member who asks "why did you pause the contract review use case?" gets the Workflow Reality Map findings and the fit test scores for that use case. An acquirer who asks "is the Scorecard methodology consistent across all use cases?" gets the Scorecard with the scoring criteria and the owner assignments for each entry.

Producing support documents only when asked has two important effects. It keeps the primary conversation focused on the Brief rather than getting lost in methodology. And it signals to the audience that the documentation exists and is available, which is itself a credibility signal. A leader who can immediately produce the supporting document for any question asked in a financial review is demonstrating that the Brief is the top layer of a complete, organized evidence base, not a polished summary of guesswork. That demonstration of depth, even when most of the depth never gets examined, builds confidence in the Brief and in the leadership team presenting it.

One common mistake in preparing the support packet is including documents that aren't directly referenced by any number in the Brief. The support packet should contain only documents that can be directly traced to a specific number, a specific decision, or a specific claim in the Brief. If a document in the support packet doesn't connect to anything in the Brief, it doesn't belong in the packet. Extra documentation doesn't add credibility. It adds confusion and signals that the Brief and the evidence behind it weren't built as a connected system. Every document in the support packet should have a clear line connecting it to a specific element of the Brief, and that connection should be obvious enough that a financial stakeholder can find it without being guided through it.

For acquirer due diligence specifically, the support packet may need to be more comprehensive than it would be for a CFO or board presentation. Acquirers conducting formal due diligence will want to see the full AI operating record, including governance documentation from Volume III of this series, readiness decisions from Volume II, and the audit findings from Volume I, alongside the performance evidence from this volume. In that context, the Executive AI Efficiency Brief serves as the executive summary of the entire AI operating system, not just the performance measurement work. The Brief doesn't change in format or length. But the support packet behind it expands to include the full AI operating record that demonstrates the business has built AI discipline across all four volumes of the operating system, from initial audit through governance through readiness through performance measurement. That complete record is what makes the AI operating system a documented, defensible, enterprise-value-enhancing asset rather than a collection of tools and experiments that happens to have a clean summary document on top of it.

Building Your First Executive AI Efficiency Brief. A Step-by-Step Process

The Brief is built from the work already done in this book. Nothing in this section requires new measurement, new analysis, or new data collection. The inputs exist. The task is organizing them into the five-section format and making the document clean, specific, and ready to present to the first audience that needs to see it.

Start by identifying the audience for the first Brief. This matters before any other decision because the audience determines which sections need the most emphasis and which cover note framing is most appropriate. If the first presentation is to a CFO, the ROI results and performance metrics sections need to be especially precise and the cost denominator needs to be fully disclosed. If the first presentation is to a board, the optimization priorities and next-period commitments sections need to be especially specific and accountability-focused. Identify the audience first. Build the Brief second.

Pull the inputs for each section from the documented work in the previous chapters. For the use case summary, use the active use case list from the AI Efficiency Scorecard. For the efficiency performance metrics, use the current and baseline measurements from Chapter 6 and the Scorecard scores from Chapter 7. For the ROI results, use the completed AI ROI calculations from Chapter 8, including both the percentage and the net dollar figure for each use case. For the optimization priorities, use the decisions from the most recent Scorecard review: which use cases are being expanded, which are being refined, which are being paused, and what specific actions are attached to each decision. For the next-period commitments, identify the three to five most important measurements, improvements, or reviews that leadership will complete before the next

Brief is produced, assign a named owner to each one, and set a specific completion date.

Write the Brief in plain language. Every sentence should be readable by someone who is financially literate but not operationally involved in the AI workflows. Avoid tool names as primary identifiers. Avoid methodology language in the main text of the Brief. Avoid percentages without dollar figures. Avoid claims that can't be traced to a specific number in the underlying evidence. If a sentence in the Brief requires a footnote to be understood, rewrite the sentence. The Brief should stand on its own as a clear, direct communication of AI performance and AI decisions. The methodology belongs in the support packet, not in the Brief itself.

Review the completed Brief against one specific standard before presenting it to anyone: can every number in this document be traced to a specific documented source in the support packet? Go through the Brief line by line. For each number, ask where it came from and confirm that the source document exists in the support packet. For each claim, ask what evidence supports it and confirm that evidence is available. For each decision or recommendation, ask what data drives it and confirm that data is documented. If any number, claim, or recommendation fails this traceability check, either fix the Brief or fix the documentation before the presentation. A Brief that can't pass its own traceability check is not ready to present to a financial stakeholder who will ask exactly these questions.

Once the Brief passes the traceability check, prepare the support packet using the six documents described in the previous section. Organize the support packet so that each document is clearly labeled and easy to find when a specific question requires it. A CFO who asks about the rework calculation should be able to receive the relevant support document within thirty seconds of asking. If the support packet is disorganized, the delay in finding the right document creates the impression that the evidence is incomplete, even when it isn't. Organization of the support packet is a credibility detail, and credibility details matter in high-stakes financial conversations.

Schedule the presentation. Don't wait for the perfect moment. The Brief is ready when it passes the traceability check and the support packet is organized.

The perfect moment for a CFO, board, lender, or acquirer conversation is the moment when the business has a clean, defensible document to present. That moment is now. Waiting for more data, more precision, or more time to refine the Brief is the most common reason strong AI measurement work never reaches the stakeholders who need to see it. A Brief built on honest, documented evidence is ready to present. Present it.

A Hypothetical Scenario: What the Brief Looks Like in Practice

To make the Brief concrete, consider a hypothetical scenario that captures how this tool functions in a real leadership context. This is a composite example, not a real case, but it reflects the kind of situation many leadership teams are navigating right now.

Return to the 60-person professional services firm. Its COO has been guiding AI use across three workflows for about fourteen months: client proposal generation, internal project status reporting, and invoice processing. She's been building toward this chapter's work for the past three months, mapping workflows, running fit tests, building the Scorecard, and completing the ROI calculations. Her CFO has a budget review in three weeks where AI spend will be on the agenda. Her board meets in six weeks and has asked for an AI performance update. She has the data. She needs the Brief.

She builds the master Brief with all five sections fully populated. The use case summary names all three workflows, explains the operational purpose of each, and states the team that uses each one. The efficiency performance metrics show that the proposal generation workflow has reduced average cycle time by twenty-two percent compared to baseline, with first-pass accuracy improving from sixty-eight percent to eighty-one percent over the past quarter. The project reporting workflow is showing flat cycle time with a declining error rate, a mixed result that the Scorecard flagged as needing attention. The invoice processing workflow is showing a fourteen percent capacity increase with stable error rates, a clear improvement. The ROI results show a forty-three percent ROI on the proposal workflow, a negative eleven percent ROI on the project reporting workflow, and a thirty-one percent ROI on the invoice processing workflow. The optimization priorities section names a refine decision on the project reporting workflow, with a specific plan to rewrite the prompting standards and redesign the review checkpoint, owned by the operations director, with a completion date six weeks out. The next-period commitments include re-scoring the project reporting use case on the fit test in sixty days, completing the prompting standards revision in thirty days, and producing the next Brief before the board meeting in six weeks.

For the CFO meeting, she writes a one-paragraph cover note that directs the CFO's attention to the ROI results section first, notes that the cost denominator includes the full AI Efficiency Tax for all three workflows, and flags the negative ROI on the project reporting workflow as the primary focus of the optimization work in progress. The CFO reviews the Brief in four minutes, asks two questions about the cost denominator methodology, receives the ROI calculation worksheets from the support packet, and approves the continued investment in the proposal and invoice workflows while asking for a sixty-day update on the project reporting refine decision. The meeting takes twenty-two minutes. Every question gets answered with a document. Every decision is made on evidence.

For the board meeting, she uses the same Brief with a different cover note that directs the board's attention to the optimization priorities and next-period commitments sections, framing the negative ROI on the project reporting workflow as evidence that leadership is actively testing and evaluating AI use cases rather than funding them indefinitely without accountability. The board's questions focus on the governance process behind the refine decision and the timeline for re-measurement. Both questions are answered directly from the Brief and the decision log in the support packet. The board leaves the meeting with confidence that AI is being managed with discipline. The COO leaves with a clear mandate to complete the project reporting optimization work and return with results at the next meeting.

The First Brief Is Always Wrong: Iterating Through Three Cycles

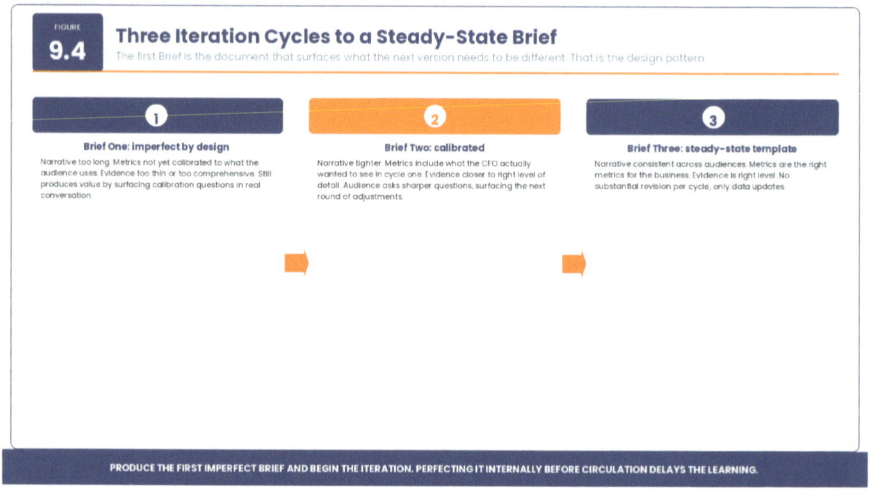

The first Executive AI Efficiency Brief a leadership team produces is going to be imperfect. This is not a failure of the leadership team or the methodology. It is the normal pattern. Briefs are operational documents that improve through iteration, and the first version of any operational document is the one that surfaces what needs to be different in the next version. (Burgess, 2018) Leadership teams that expect the first Brief to be polished often delay producing one at all, which means they never start the iteration cycle that actually produces a strong Brief. The faster path is to produce a first Brief that is acknowledged as imperfect, then improve it in each subsequent cycle.

The first Brief typically has three patterns of imperfection. The narrative is too long, the metrics are not yet calibrated to what the audience actually wants to see, and the supporting evidence is either too thin or too comprehensive. The narrative is long because the team has not yet learned what to include and what to leave for the supporting documents. The metrics are not calibrated because the team has not yet had the conversation with the CFO or the board about what they actually use to assess AI performance. The evidence is misjudged in both directions because the team has not yet seen which questions the audience will ask. The first Brief still produces value, because it is the document that surfaces these calibration questions in the actual operating conversation rather than in the team's internal review.

The second Brief, produced one quarter later, addresses the calibration issues the first Brief surfaced. The narrative is tighter. The metrics include the ones the CFO actually wanted to see in the first conversation. The supporting evidence is closer to the right level of detail. The audience for the second Brief typically responds with sharper questions than they did to the first one, because they now have a reference frame from the prior quarter. Those sharper questions surface the next round of calibration adjustments, which feed into the third Brief.

The third Brief, produced two quarters after the first, is usually the version that becomes the steady-state template for the AI program. The narrative is consistent in length and structure across audiences. The metrics are the right metrics for the business. The supporting evidence is at the right level for the questions the audience typically asks. The Brief no longer requires substantial revision each cycle, only updates to the underlying data. From this point forward, the Brief becomes a routine operating instrument, and the leadership team's attention can shift from producing the Brief to using it as the entry point for the operating decisions it supports.

The iteration is not a sign of weakness. It is the design pattern. Volume III's per-use-case governance dossiers had the same iteration cycle. Volume II's workflow readiness records had the same cycle. Volume I's audit findings had the same cycle. Operating documents improve through use (Khalil et al., 2025) , and the first version exists primarily to surface what the second version needs to be different. Leadership teams that internalize this pattern produce strong Briefs faster than teams that try to perfect the first version before circulating it.

When to Update the Brief Mid-Quarter

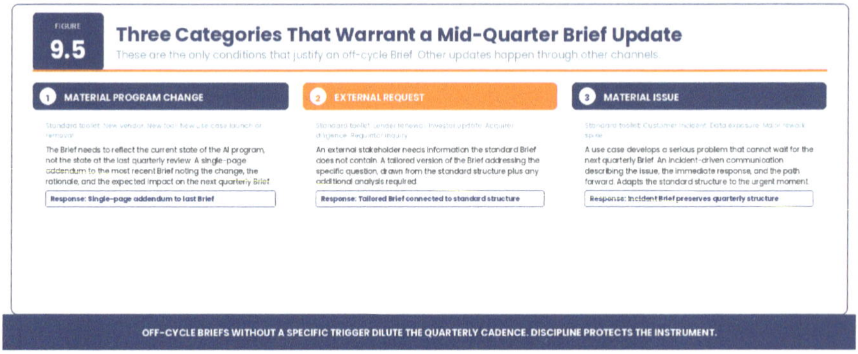

The standard Brief cadence is quarterly, aligned to the Scorecard review cycle. There are operating moments when waiting for the next quarterly cycle is wrong, and the Brief needs to be updated mid-quarter to support a specific decision. Recognizing these moments is part of the operating discipline, and the protocol for mid-quarter updates determines whether the Brief remains a useful instrument or becomes a routine that the team is producing for its own sake.

Three categories of mid-quarter event typically warrant an update. The first is a material change in the AI program itself: a new vendor relationship, a major tool change, the launch of a new use case, or the removal of an existing one. The Brief needs to reflect the current state of the program rather than the state at the last quarterly review. The mid-quarter update is usually focused: a single-page addendum to the most recent Brief, noting the change, the rationale, and the expected impact on the next quarterly Brief. The addendum is sent to the same audience that received the last Brief.

The second category is an external request that the standard Brief does not address. A lender preparing for a credit facility renewal, an investor preparing for a quarterly update, an acquirer running diligence, or a regulator opening an inquiry may need information that the standard Brief does not contain. The mid-quarter response is typically a tailored version of the Brief addressing the specific question, with content drawn from the standard Brief plus any additional analysis the request requires. The discipline is to keep the tailored version connected to the standard Brief structure, so that the external audience sees a coherent operating record rather than a one-off document.

The third category is the surfacing of a material issue between Briefs. A use case that develops a serious problem (a customer incident, a data exposure issue, a major rework cost spike) cannot wait until the next quarterly Brief for leadership communication. The mid-quarter update in this case is incident-driven: a focused communication describing the issue, the immediate response, and the path forward. The standard Brief structure adapts to this purpose by treating the incident as the headline and the regular content as supporting context. The mid-quarter incident Brief preserves the structural integrity of the quarterly Brief while addressing the urgent moment.

The risk to avoid is the mid-quarter Brief that has no specific operating purpose. Briefs produced because the team feels they should produce one, without a triggering event or a specific audience need, dilute the quarterly cadence and erode the Brief's status as a meaningful decision document. The discipline is to produce mid-quarter Briefs only when one of the three categories above applies, and to handle other communication needs through other channels (operating reviews, governance meetings, ad-hoc updates) that do not weaken the Brief itself.

That's what the Executive AI Efficiency Brief produces when it's built correctly and presented with the right support structure behind it. Not applause for an impressive AI story. Confidence in a disciplined AI operating system. And confidence, earned through evidence, is the outcome that makes every subsequent AI conversation faster, cleaner, and more productive than the one before it.

Putting It Into Practice

This chapter built the Executive AI Efficiency Brief as the communication instrument that converts the Scorecard scores, the ROI calculations, and the optimization priorities into the format a financial stakeholder can actually use. The chapter's work is not finished when the structure is understood. It is finished when the first Brief sits in front of a CFO, board, lender, or acquirer with every number traceable to a documented source. Before turning the page, complete the four steps below.

First, populate the five sections of the Brief. Use case summary, performance metrics, ROI results, optimization priorities, next-period commitments. One page total. Pull the numbers from the Chapter 7 Scorecard and the Chapter 8 ROI calculations. Every sentence in the Brief should be able to point to a specific document behind it. If a claim in the Brief cannot trace to a document in the supporting packet, the claim does not belong in the Brief.

Second, tailor the cover note for the chosen audience. The five sections do not change between audiences. The framing does. A CFO reads numbers first and the story second, so the cover note leads with ROI and the cost denominator. A board focuses on priorities and commitments. A lender wants to see consistency and cost discipline. An acquirer wants to see depth of documentation. Pick the first audience the Brief will reach and write the cover note to that audience specifically, not to a generic financial reader.

Third, assemble the supporting documentation packet. Six items minimum: the Scorecard, the ROI worksheets per use case, the baseline measurements from Chapter 6, the Fit Test results from Chapter 5, the Reality Maps from Chapter 4 for any use case receiving a pause or remove decision, and the decision log. The packet stays in the room with the Brief but does not lead the conversation. Produce specific documents only when the audience asks specific questions that require them.

Fourth, calendar the next iteration before the first Brief leaves the room. The first Brief is wrong in specific ways that only become visible when the audience reads it. The narrative is too long or too short, the metrics are not yet the ones the CFO actually uses, the evidence is either thin or comprehensive in the wrong proportions. Capture the questions the audience asks and the gaps they surface. Those become the calibration inputs for the second Brief. Three iterations produce the steady-state template that holds across audiences.

The Brief is one page. The supporting evidence sits behind it. The leadership team that walks into a board meeting, a lender review, or an acquisition conversation with a current Brief in hand has done something most leadership teams in this position have not done: they have prepared. The next chapter addresses the work that keeps the Brief honest from one quarter to the next. Consistent AI outputs across the team. Consistent measurements from one review to the next. A standardization discipline that means the Brief in quarter four reflects the same operating reality the Brief in quarter one reported, with every quarter showing measurable movement rather than measurement noise. Standardization is what holds the Brief together over time, and it is the work this Volume turns to next.

Chapter 10: Standardizing AI Across the Team

Eliminating Shadow Processes and Inconsistent Usage

CHAPTER 10 OUTCOMES

At the end of this chapter, you will be able to:

1. Run the six-question structured conversation that surfaces shadow AI processes in thirty minutes per workflow.
2. Classify each shadow process finding as a better method, a workaround for a broken step, or an unapproved risk practice.
3. Draft an AI usage standard for any workflow using the eight-question framework that fits on one page.
4. Lead the five-component leadership alignment conversation before standards reach the team.
5. Build the standards library as a permanent operating asset that survives team changes and ownership transitions.

STANDARDIZATION IS NOT A POLICY EXERCISE. IT IS AN OPERATING DISCIPLINE THAT MAKES AI USAGE REPEATABLE.

The Problem This Chapter Solves: Everyone on Your Team Is Using AI Differently and It's Costing You

By the time a business has mapped its workflows, calculated its AI Efficiency Tax, built its Scorecard, and produced a defensible ROI number, it has accomplished something most organizations never reach. It has real evidence. Real measurements. Real decisions attached to real use cases. That's the operating foundation this entire book has been building toward.

But there's a problem that sits underneath all of that work, and it's one that measurement alone can't fix.

When ten people use the same AI tool in ten different ways, the business doesn't have a standardized AI operation. It has ten separate experiments running simultaneously inside the same workflow. Each person may be using different prompts, different source material, different quality judgments, and different review habits. Some employees have built personal prompt libraries that produce excellent results. Others are running the same tool with minimal context and producing outputs that require significant correction. A few may have quietly stopped using the approved tool altogether and switched to something they found on their own. And nobody in leadership knows exactly which version of the workflow is actually running on any given day.

That inconsistency has a compounding cost (Lee et al., 2025) . It shows up as rework when a manager reviews outputs from two different team members and finds they're structured completely differently. It shows up as review burden when a senior employee has to spend extra time checking AI outputs because quality varies so much across the team that no output can be trusted without inspection. It shows up as client confusion when deliverables look inconsistent from one engagement to the next. And it shows up as management overhead when leaders spend time resolving quality disputes that trace back to the fact that nobody agreed on how the tool was supposed to be used in the first place.

This is an operating problem, not a training problem (Webber et al., 2019) . Training might teach employees how to use the tool. It doesn't create a shared standard for how the tool should be used inside a specific workflow, what the output should look like, when human review is required, and what happens when the AI output isn't good enough to use. Without those standards, every employee is making those judgments individually, every day, and the business is paying for the inconsistency that results.

Leadership needs a practical way to bring AI usage out of scattered individual habits and into a shared operating model the team can actually follow. Not by micromanaging every interaction with an AI tool, but by creating enough structure that AI-supported work becomes repeatable, measurable, and reliable across the entire team rather than dependent on which employee happened to run the workflow that day.

How to Identify Shadow AI Processes and Undocumented Workarounds

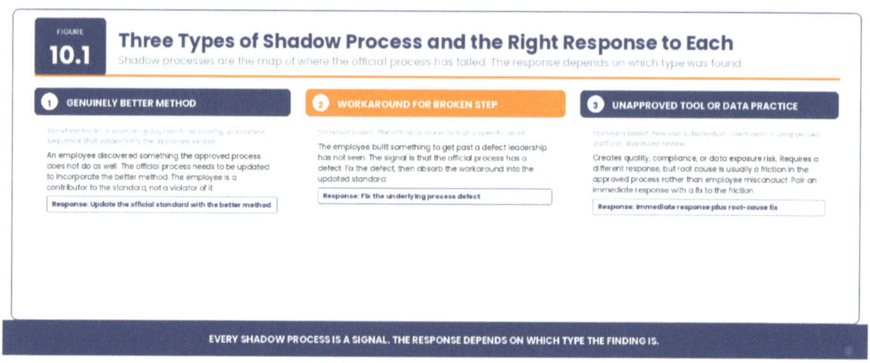

Shadow AI is not a new topic in this Library. Volume I Chapter 2 is the foundational treatment: what shadow AI is, why employees adopt it, and how to surface it through a discovery exercise rather than a compliance audit. Volume II Chapter 7 introduces the AI Adoption Pattern Map and scores shadow AI on the AI Readiness Maturity Scale as an internal control issue. Volume III Chapter 4 quantifies the cost: in the IBM and Ponemon 2025 data, shadow AI added roughly USD 200,000 to the average breach and extended the breach lifecycle by six days. This chapter does not repeat that material. It adds the layers Volumes I, II, and III did not produce: a monthly dollar figure on shadow process cost (carried in the AI Efficiency Tax from Chapter 3), a three-type taxonomy with differentiated responses, an inline workflow signal during Reality Map work (Chapter 4), and a recurring shadow process incidence rate on the AI Efficiency Scorecard (Chapter 7). The Library's shadow AI treatment is now four-Volume layered: surface (Volume I), diagnose (Volume II), quantify breach risk (Volume III), measure ongoing operating cost and integrate into quarterly cadence (this Volume).

Shadow AI processes are the informal ways employees use AI outside the documented process (Semionovas, 2026) , the approved tool list, or the measurement system. They aren't usually the result of bad intentions (Silic et al., 2025) . They form because the official process is too slow, too unclear, too disconnected from how the work actually gets done, or simply not good enough to produce the results the employee needs. When the approved process fails the employee, the employee builds something that works. That something is the shadow process.

Shadow processes are almost always invisible to leadership until someone goes looking for them specifically (Vale, 2025) . They don't appear in usage reports. They don't show up in workflow documentation. They live in personal prompt files saved on individual desktops, in informal tool subscriptions paid out of pocket or expensed without a category, in undocumented steps employees have added to the workflow because experience taught them the official version was unreliable, and in workarounds that have become so routine nobody thinks of them as workarounds anymore.

The detection method that works is a direct, structured conversation with the people doing the work (Prati et al., 2003) . Not a survey. Not an audit. A conversation, conducted in a way that makes it safe to be honest.

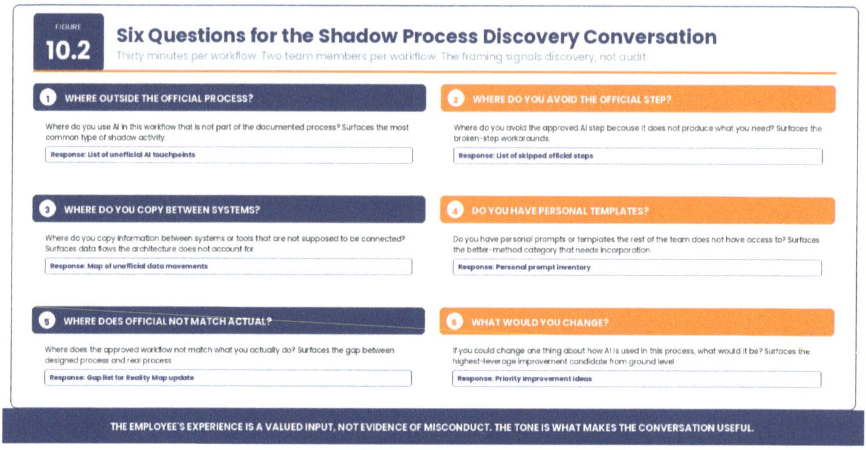

Start by identifying the five to eight workflows in the business where AI is most active. For each workflow, schedule a thirty-minute conversation with one or two of the team members who execute it daily. Ask six specific questions. First, where do you use AI in this workflow that isn't part of the official process? Second, where do you avoid the official AI step because it doesn't produce what you need? Third, where do you copy information between systems or tools that aren't supposed to be connected? Fourth, do you have personal prompts or templates you use that the rest of the team doesn't have access to? Fifth, where does the approved workflow not match what you actually do? Sixth, if you could change one thing about how AI is used in this process, what would it be?

Those six questions will surface more shadow process information in thirty minutes than a month of usage report analysis, provided the tone follows the discovery posture Volume I Chapter 2 established for shadow AI conversations more broadly.

The framing that works is direct and specific. Tell the team member exactly why you're having the conversation: the business is redesigning how AI is used in this workflow, and the only way to build something better than what currently exists is to understand what's actually happening right now. That framing signals that the employee's experience and knowledge are valued inputs, not evidence of misconduct. It also signals that something is going to change, which makes the conversation feel useful rather than threatening.

What you'll find in most businesses that have been using AI for more than a year is a mix of three types of shadow processes. The first type is a genuinely better method. An employee discovered a prompting approach, a tool configuration, or a workflow sequence that produces significantly better results than the approved version. This is valuable information. It means the official process needs to be updated, not that the employee needs to be corrected. The second type is a workaround for a broken step. The official process fails at a specific point and the employee built something to get past it. This is a signal that the official process has a defect that leadership hasn't seen. The third type is an unapproved tool or data practice that creates risk. An employee is using a personal AI subscription, feeding client data into an unapproved platform, or bypassing a review step in a way that creates quality or compliance exposure. This type requires a different response, but even here the root cause is usually a failure in the official process rather than an employee acting irresponsibly.

Consider how this works in practice. Return to the 60-person professional services firm. Its COO has been concerned for months that AI usage across the consulting teams varies more than anyone can describe. She schedules thirty-minute conversations with two consultants from each of three workflows: client deliverable drafting, research synthesis, and weekly status reporting. Six conversations across one week surface a clear pattern. Three of the consultants on the deliverable workflow use personal prompt libraries they built outside the approved tool because the official prompts produce generic output. One consultant on the research workflow pastes client data into a free AI tool on a personal laptop to avoid what the team calls "the slow approval queue." Both consultants on the status-reporting workflow have stopped using the approved template altogether because they found a faster manual approach. The findings sort cleanly into the three types. The personal prompt libraries are genuinely better methods that need to be incorporated into the official standard. The skipped status-report template is a workaround for a broken step that needs the underlying process fixed. The free AI tool with client data is the higher-risk practice that requires an immediate response paired with a fix to the friction in the approved tool. Three workflows. Six conversations. Three patterns. The standardization queue writes itself.

Document every shadow process finding by workflow, by type, and by the specific step in the process where it occurs. That documentation becomes the input for the standardization work in the next section. The shadow processes aren't the problem to eliminate. They're the map of where the official process has failed, and that map is exactly what's needed to build something better.

Building AI Usage Standards That Teams Actually Follow

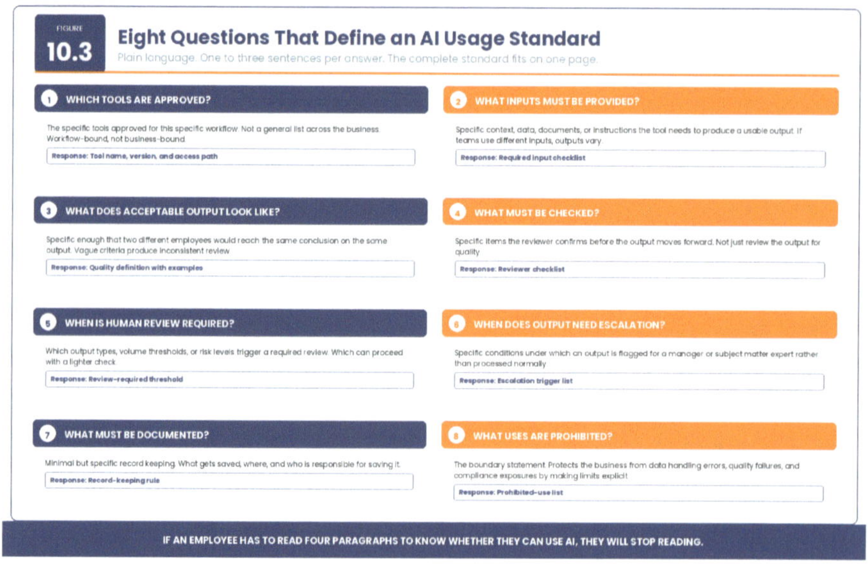

An AI usage standard only works if it fits the workflow people actually perform (Retelny et al., 2017) . A standard that describes how the process was designed to work, rather than how it actually works, will be ignored within weeks of publication. A standard that adds so many checkboxes, approvals, and documentation requirements that using AI takes longer than not using it will generate its own shadow process within months. The goal is a standard that makes the right way to use AI the easiest way to use AI (Arora & Baenziger, 1986) , so that following it requires less effort than working around it.

That goal requires a specific design approach. The standard has to be built from the workflow up, not from a compliance document down.

Start with the workflow findings from Chapter 4 and the shadow process inventory from the previous section. These two inputs tell you what the workflow actually looks like and where the current AI usage is producing inconsistent results. The standard you build should address both. It should reflect the real workflow, not the designed one, and it should resolve the specific inconsistencies the shadow process review surfaced.

Every AI usage standard needs to answer eight specific questions for the team member using it.

First, which AI tool or tools are approved for this workflow? Not a general list of approved tools across the business, but the specific tools approved for this specific workflow.

Second, what inputs must be provided to the AI tool before it runs? This means the specific context, data, documents, or instructions the tool needs to produce a usable output. If the team is using different inputs, the outputs will vary, and the standard must define what consistent input looks like.

Third, what does an acceptable output look like? This is the quality standard. It should be specific enough that two different employees reviewing the same AI output would reach the same conclusion about whether it meets the standard. Vague quality criteria produce inconsistent review decisions.

Fourth, what must be checked before the output moves forward? This is the review requirement. It should list the specific things a reviewer must confirm, not just "review the output for quality."

Fifth, when is human review required and when can the output move forward without it? Not every AI output needs a senior review. The standard should define which output types, volume thresholds, or risk levels trigger a required review and which ones can proceed with a lighter check.

Sixth, when does the output need escalation? This is the exception handling rule. It defines the specific conditions under which an output should be flagged for a manager or subject matter expert rather than processed through the normal workflow.

Seventh, what must be documented about the AI interaction? This is the record-keeping requirement. It should be minimal but specific. What gets saved, where it gets saved, and who is responsible for saving it.

Eighth, what uses of AI are prohibited in this workflow? This is the boundary statement. It protects the business from data handling errors, quality failures, and compliance exposures by making the limits explicit rather than leaving them to individual judgment.

Those eight questions define the complete standard for any AI-supported workflow. Write the answers in plain language. Keep each answer to one to three sentences. The standard for a single workflow should fit on one page. If it doesn't fit on one page, it's too complex to follow during real work, and it will be ignored.

The design rule that separates a usable standard from a decorative one is this: every standard should answer the three questions that employees actually ask themselves when using AI at work. Can I use AI here? What do I need to check before I use the output? When do I need to get someone else involved? If the standard answers those three questions clearly and quickly, it will be used. If an employee has to read four paragraphs to figure out whether they're allowed to use AI for a specific task, they'll stop reading and make their own judgment. The standard should make the right answer to each of those three questions immediately obvious.

Once the standard is drafted, test it with the team members who will use it before finalizing it. Ask them to walk through a recent AI interaction using the standard as their guide. Where they hesitate, the standard is unclear. Where they skip a step, the standard is too burdensome. Where they say "that's not how it actually works," the standard doesn't match the real workflow. Revise based on what the test reveals. A standard that survives one test run with a real team member is significantly more likely to be followed than one that was written entirely from above without any ground-level input.

The final version of the standard should be stored in one place that every team member who uses that workflow can access immediately. Not in a shared drive folder that requires three clicks to find. Not in a policy document that nobody reads. In the same location where the workflow instructions live, attached to the workflow itself so that the standard and the process are inseparable from the team member's perspective.

FIGURE 10.4 **Five-Component Framework for the Leadership Alignment Conversation**

Hold this conversation before standards reach the team. Each component addresses a different source of misalignment

ALIGNMENT BEFORE ROLLOUT. WITHOUT IT, FRAGMENTED LEADERSHIP SIGNALS PRODUCE EXACTLY THE INCONSISTENCY STANDARDIZATION WAS DESIGNED TO ELIMINATE.

Getting Leadership Team Alignment Around AI Standardization

AI standardization will fail if the leadership team isn't aligned before the standards reach employees. That's not a political observation. It's an operating reality. When a CFO, a COO, and three department heads are all approaching AI standardization with different priorities and different concerns, the message that reaches the team is fragmented. Employees receive conflicting signals about how seriously to take the new standards, whether exceptions will be granted, and whether leadership is unified behind the change or quietly divided on it. That fragmentation produces exactly the inconsistency the standards were designed to eliminate.

The alignment conversation needs to happen before rollout, not during it.

The most common tension in this conversation is between financial accountability and operational flexibility. A CFO focused on cost and measurable return wants standards tight enough to produce consistent, measurable results. A COO focused on workflow speed wants standards light enough that they don't slow execution. Department heads focused on local needs want enough flexibility to handle the specific requirements of their team's work without being forced into a one-size-fits-all process. These are legitimate concerns, and they don't resolve themselves without a structured conversation that names them directly.

The framework for that conversation has five components. Work through them in order. Don't skip any of them, because each one addresses a different source of misalignment and leaving any of them unresolved will surface as a problem during implementation.

The first component is purpose. What is the business trying to achieve through AI standardization? The answer should be specific: consistent output quality, measurable efficiency gains, reduced review burden, defensible ROI, and an AI operating model that can be improved over time. Not "better AI usage" or "more consistent processes." Specific outcomes that connect to the financial results the leadership team is accountable for. When everyone agrees on the purpose, disagreements about specific standards become easier to resolve because there's a shared test: does this standard support the agreed purpose or not?

The second component is scope. Which workflows will be standardized first? The answer should be the three to five workflows where AI activity is highest, AI Efficiency Tax is greatest, or inconsistency is creating the most downstream cost. Starting with a focused scope keeps the initial standardization effort manageable and produces early results that build organizational confidence in the process.

Trying to standardize every AI use case simultaneously is a recipe for an incomplete rollout that loses momentum before it produces results.

The third component is minimum rules. What are the non-negotiable standards that every team must follow, regardless of their specific workflow or department? These are the floor, not the ceiling. They might include a requirement that every AI output used in a client-facing deliverable must pass a documented review, a prohibition on feeding personally identifiable information into unapproved tools, or a requirement that every team using AI for a specific workflow category must use an approved prompt template. Minimum rules protect the business from the highest-risk inconsistencies without constraining the flexibility that department heads need to manage their specific workflows effectively.

The fourth component is metrics. What will leadership measure to confirm that standardization is working? The answer should connect directly to the AI Efficiency Scorecard from Chapter 7. Cycle time consistency across team members doing the same AI-assisted task. First-pass accuracy rates before and after standards are implemented. Rework frequency per workflow. Shadow process incidence rate, meaning how often team members report using an approach outside the approved standard. These metrics give leadership a way to confirm that standardization is producing the intended results rather than just generating documentation that nobody follows.

The fifth component is enforcement ownership. Who is responsible for confirming that standards are being followed, identifying when they're not, and escalating when a team member or department is consistently operating outside the approved process? This doesn't need to be a compliance function or a surveillance role. It can be the workflow owner identified in the AI Efficiency Scorecard, the team lead for each workflow, or the operations leader responsible for the department. What matters is that someone is named, that the responsibility is explicit, and that the person named has the authority and the information needed to act when standards are being ignored.

Alignment doesn't require everyone in the room to be equally enthusiastic about AI standardization. A CFO who is skeptical about AI in general can still agree on the purpose, the scope, the minimum rules, the metrics, and the enforcement ownership. A department head who values flexibility can still agree on a minimum rule set that protects the business without constraining how their team handles workflow-specific variations. The goal of the alignment conversation isn't consensus on AI enthusiasm. It's agreement on how AI will be used, measured, reviewed, and corrected. That's a narrower and more achievable goal, and it's the only one that matters for standardization to work.

The leadership message that follows this alignment conversation should be consistent across every department and every level of the organization. Standardization isn't about slowing AI down or reducing the team's flexibility to get work done. It's about making AI reliable enough to scale. A team using AI in ten different ways produces ten different quality levels, ten different review burdens, and ten different risk profiles. A team using AI within a shared standard produces consistent quality, predictable review burden, and a measurable result that leadership can improve over time. That's the message. Say it the same way every time it comes up.

Maintaining Standards Without Creating Bureaucracy

The biggest risk after AI standards are created isn't that employees will ignore them, though some will. The biggest risk is that the standards become a document nobody uses because the process for maintaining them is heavier than the process for working around them.

Standards decay in predictable ways. A tool gets updated and the prompting guidance in the standard no longer produces the results it used to. A workflow changes and the review requirement in the standard no longer matches the actual review step. An employee discovers a better approach and uses it without updating the standard, so the standard now describes a suboptimal process that nobody follows. The business adds a new use case and nobody creates a standard for it because the process for creating standards feels too burdensome to initiate for every new workflow. Within six to twelve months, the standards that were carefully built are partially outdated, partially ignored, and partially replaced by informal practices that have quietly become the new shadow processes.

The maintenance model that prevents this is light, recurring, and built into the operating rhythm rather than treated as a separate project.

A quarterly standard review is the minimum cadence. It should take no more than sixty minutes per workflow and should cover five specific questions. First, is the approved tool still the right tool for this workflow, or has a better option become available? Second, are the prompting guidelines still producing consistent, high-quality output, or have they drifted out of alignment with the tool's current capabilities? Third, has the review requirement changed based on the error rate and rework data from the AI Efficiency Scorecard? If first-pass accuracy has improved significantly, the review requirement may be reducible. If it has declined, the review requirement may need to be strengthened. Fourth, are there any shadow processes that have appeared since the last review, and if so, do they represent a better method that should be incorporated into the standard? Fifth, who currently owns this standard, and is that ownership still appropriate given any changes in team structure, workflow volume, or business priorities?

Those five questions can be answered in a structured conversation with the workflow owner and one or two team members. The output is a brief update note that documents what changed in the standard, why it changed, who approved the change, and when the updated standard was communicated to the team. That note becomes part of the AI operating record and demonstrates to any stakeholder reviewing the governance documentation that standards are actively maintained rather than created once and abandoned.

Spot checks are a lightweight complement to the quarterly review. A spot check is a brief, informal look at a sample of AI outputs from a specific workflow to confirm that the quality, format, and review practices described in the standard are actually being followed. It doesn't require a formal audit process. A workflow owner reviewing five to ten outputs per month and noting whether they meet the standard's quality criteria is sufficient for most SMB workflows. The goal isn't to catch employees doing something wrong. It's to detect early when a standard has drifted out of alignment with how the work is actually being done, before that drift becomes expensive.

Shared prompt templates and output quality checklists deserve specific attention in the maintenance model. These are the operational tools that make standards usable on a daily basis, and they require more frequent updates than the standard itself. When a tool update changes how prompts behave, the shared templates need to be updated before the team continues using them. When a quality problem appears in the spot check data, the output quality checklist should be updated to include the specific check that would have caught that problem. Assigning one person per workflow the explicit responsibility for keeping the templates and checklists current is the simplest and most effective way to prevent these operational tools from going stale.

The update process for standards should be as simple as possible. When a team member discovers a better method, there should be a clear, low-friction path for proposing that method as a standard update. A brief written description of the new approach, the evidence that it produces better results, and a recommendation to the workflow owner is sufficient. The workflow owner reviews it, tests it if needed, and either incorporates it into the standard or explains why the current approach is being maintained. The whole process should take less than a week for routine updates. A standard that can only be updated through a formal change management process will never be updated, and it will be replaced by shadow processes within the year.

The principle that ties all of this together is straightforward. The best AI operating discipline learns from the team rather than freezing the process in place. Employees who use AI tools every day will discover better approaches, identify new failure modes, and develop judgment about when AI works and when it doesn't that no standard written from above can fully capture. A maintenance model that incorporates those discoveries into the official standard converts individual learning into organizational capability. It makes the standard better over time rather than letting it become increasingly irrelevant as the team's real practices diverge from the documented ones.

How to Identify the Right Standards to Document First

Not every AI usage pattern in the business needs to be standardized, and trying to standardize everything at once produces documentation that the team will not follow and standards that the leadership will not maintain. The discipline is to triage the standardization opportunities by impact and tackle the highest-impact ones first, leaving the lower-impact ones for later cycles or accepting that some patterns are not worth formal standardization at all. The triage methodology runs on three questions applied to each candidate standard.

The first question is whether inconsistent usage of this pattern is producing measurable cost. Patterns that the team handles inconsistently but where the inconsistency does not show up in cycle time, error rate, rework, or customer outcomes are not worth standardizing first. The labor invested in producing and maintaining a standard for a pattern that is not producing measurable cost is labor that is not producing measurable benefit. Standards that earn their place demonstrate, in the measurement record, that inconsistent handling of this pattern is contributing to the AI Efficiency Tax in some form. The Tax connection is the practical test for whether the standard belongs at the top of the standardization queue.

The second question is whether the team would actually follow this standard if it existed. Some patterns are inconsistent because the team has not received a clear standard. Others are inconsistent because the work genuinely varies across cases and a one-size-fits-all standard would not fit. Patterns in the first category are good candidates for standardization. Patterns in the second category are not, even if the inconsistency is producing measurable cost, because the standard would either be ignored or produce its own problems through over-specification. The diagnostic is to ask the team directly whether a standard would help or hinder their work in this pattern. The team's honest answer is the most useful input the leadership team has.

The third question is whether the standard would survive turnover. A standard that depends on one team member's interpretation will not survive that team member leaving the role. A standard that can be read, understood, and applied by a new hire after a brief onboarding session will. The test for survivability is to ask whether a new team member with relevant background but no specific knowledge of the business could read the standard and produce acceptable work. If the answer is yes, the standard is worth the documentation effort. If the answer is no, the standard either needs more thought or is not yet ready to be a formal standard.

Standards that pass all three tests go to the top of the queue. Standards that pass two of three may go to the queue as second priority. Standards that pass only one or none are not yet ready to be formal standards, and the team's effort is better spent on the higher-impact candidates. Most leadership teams find that this triage produces a queue of five to ten standards worth documenting in the first six months of the Volume IV work, with additional standards emerging as the operating discipline matures. The compressed queue is more useful than the comprehensive one because the team can actually maintain it.

Standardization in Distributed Teams: Remote-First Considerations

Teams that work in shared physical space develop informal AI usage standards through proximity. They see what their colleagues are doing, ask questions in passing, and converge on shared practices without explicit documentation. Distributed teams do not have this informal convergence mechanism. AI usage in a remote-first team often varies more widely than usage in a co-located team, because each team member is developing their own AI practice in isolation and the cross-team learning that happens in shared space does not happen in distributed work. The standardization discipline has to substitute formal mechanisms for the informal convergence that distance removes.

The first adaptation is documentation that does not assume shared context. A standard written for a co-located team can rely on shared understanding of the work context. A standard written for a distributed team has to make the context explicit. The standard includes not just the prompting pattern but the situations the pattern applies to, the inputs it expects, and the review steps that should follow. Distributed standards are typically longer than co-located standards for the same workflow, because they include context that proximity would otherwise convey.

The second adaptation is visible examples. A co-located team can show new team members examples of good AI usage through over-the-shoulder demonstration. A distributed team needs the examples in the documentation itself. Standards for distributed teams typically include three to five worked examples of the pattern applied to representative cases, with the prompts, the AI outputs, and the team's final outputs all preserved. The examples take more effort to produce than the standard itself, but they are what makes the standard usable for a team that cannot watch a colleague apply it.

The third adaptation is the standards review cadence. Co-located teams update their informal standards through ongoing conversation. Distributed teams update formal standards through a scheduled cadence. The defensible cadence for distributed teams is monthly standards reviews in the first six months of any new standard's life, then quarterly reviews thereafter. The monthly cadence catches early friction with the standard while the team is still developing fluency. The quarterly cadence maintains the standard once the team has internalized it.

The fourth adaptation is the channel for asking questions about a standard. Co-located teams ask questions across desks. Distributed teams need an explicit channel: a Slack channel, a shared document, a recurring office-hours session, or some other mechanism that lets team members raise questions about the standard without scheduling a meeting. Without the channel, distributed team members work around their questions silently, and the workaround becomes the de facto standard while the documented standard becomes the version no one is following. The channel is the operating substitute for the informal questions co-located teams handle through proximity.

The Standards Library as a Permanent Operational Asset

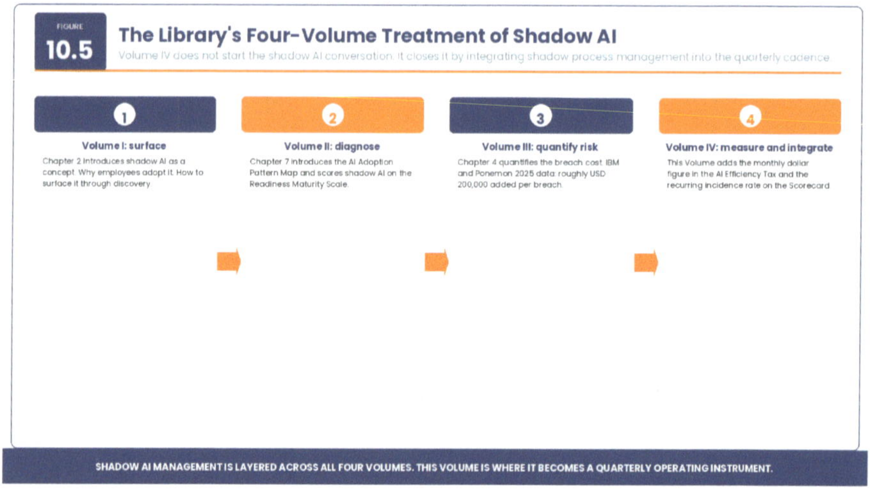

Once the leadership team has produced its first five or six standards through the triage and documentation process, the standards collectively become a library. The library is more than the sum of its individual standards. It is an operating asset that affects hiring, onboarding, knowledge transfer, and the business's resilience to turnover. Treating the library deliberately as an asset rather than as a collection of documents changes how it gets maintained and how much value it produces.

The library needs a permanent location that the team can reach without friction. A shared drive folder works. A wiki page works. A team-shared document store works. What does not work is a library that lives in the personal storage of one team member, that requires a request to access, or that is fragmented across multiple systems that team members do not know about. The library's value comes from the team's ability to reach it quickly when they need to apply a standard or when they want to confirm what the standard says. Friction in access produces team members who work without the library, which produces the inconsistent usage the library was supposed to prevent.

The library needs an index. Five or six standards can be found by browsing. Twenty or thirty standards cannot. The index is a one-page document listing every standard, its purpose, its current owner, and its last review date. The index is what makes the library searchable and what allows new team members to discover the standards relevant to their work. The index is updated whenever a standard is added, removed, or substantially revised. Without the index, the library becomes a collection of documents that the team has to remember, which is the failure mode that eventually erodes the library's usefulness.

The library needs a review cadence at the library level, not just at the individual standard level. Once a quarter, the leadership team reviews the library as a whole: which standards have been added, which have been removed, which are due for review, and what the team's overall posture toward AI usage is signaling. The review is brief, usually no more than thirty minutes, and produces a documented record of the library's state. After four quarters, the library has a developmental record that supports onboarding new team members, communicating the AI operating discipline to external stakeholders, and demonstrating the maturity of the AI program in any diligence conversation. The library becomes one of the artifacts that materially affects the business's AI valuation, which is the connection Chapter 12 examines in detail.

That's also what connects this chapter directly to the broader operating discipline this book has been building. Standardization isn't a one-time cleanup of inconsistent AI habits. It's an ongoing practice that keeps AI usage consistent, measurable, and improvable as tools evolve, workflows change, and the team's experience with AI deepens. A business that maintains its standards with the same discipline it applies to its financial reporting has an AI operating model that can be scaled, defended, and improved quarter after quarter. A business that creates standards once and never revisits them has documentation that looks like discipline from a distance and produces inconsistency from the inside. The difference between those two outcomes is the maintenance model, and the maintenance model is what this section is designed to install.

Putting It Into Practice

The work in this chapter is complete when the business has a shadow process inventory, a set of drafted usage standards for the highest-priority workflows, a leadership alignment record, and a maintenance cadence with named owners and scheduled review dates. Those four outputs are the minimum viable version of a standardized AI operating model, and they're achievable within thirty days for most SMB leadership teams.

Here are the specific steps to complete before moving into the 90 Day AI Process Optimization Plan in the final chapter.

First, conduct a shadow process review across the three departments or workflows where AI activity is highest. Use the six-question conversation format described in this chapter. Schedule thirty-minute conversations with one or two team members per workflow. Document every shadow process finding by workflow, by type, and by the specific step where it occurs. Categorize each finding as a better method, a workaround for a broken step, or an unapproved practice that creates risk. That categorization determines the appropriate response to each finding and prevents the review from being treated as a compliance exercise rather than a process improvement input.

Second, identify the three most common undocumented AI habits across the workflows you reviewed. These are the practices that appear most frequently across team members, that vary most significantly from the approved process, and that are producing the most inconsistency in output quality or review burden. These three habits are the highest-priority targets for standardization, because they're the ones generating the most cost from inconsistency right now.

Third, draft usage standards for the three highest-priority AI workflows using the eight-question framework from this chapter. Write each standard in plain language. Keep each one to a single page. Test each draft with one team member who executes the workflow daily before finalizing it. Revise based on what the test reveals. Store the final version in the same location as the workflow instructions so that the standard and the process are accessible together.

Fourth, hold the leadership alignment conversation using the five-component framework from this chapter. Work through purpose, scope, minimum rules, metrics, and enforcement ownership in order. Document the agreements reached on each component. That documentation is the alignment record that ensures the rollout message is consistent across every department head and every team lead who communicates the new standards to their teams.

Fifth, communicate the standardization rollout to the team with a specific, non-punitive message. Tell the team directly what the new standards cover, why they were created, and what problem they're designed to solve. Make it clear that the goal is consistency, quality, and measurable improvement, not surveillance or punishment for past practices. Acknowledge that some of the shadow processes the review uncovered were genuinely better methods, and that those methods have been incorporated into the new standards. That acknowledgment does two things. It validates the employees who discovered better approaches, and it signals that the standardization process is designed to learn from the team rather than override it.

Sixth, assign a named owner to each usage standard and put the first quarterly review date on the leadership calendar before the rollout is complete. The owner is responsible for the spot check process, the template and checklist updates, the quarterly review, and the escalation path when standards are being consistently ignored. The calendar date is the commitment that keeps the maintenance model from being a good intention that never gets executed.

The business that completes these six steps has done something most organizations using AI have never done. It has moved AI usage from scattered individual habits into a shared operating model that produces consistent results, measurable quality, and a defensible record of how AI is being used across the team. That operating model is what makes the 90 Day AI Process Optimization Plan in the next chapter executable rather than aspirational. Standardized AI usage is the foundation on which optimization is built, because you can't improve what you can't measure, and you can't measure what you can't standardize. The standards come first. The optimization follows. And the results, in cycle time, capacity, rework, margin, and ROI, are what the entire operating discipline is designed to produce.

FIGURE **10.6**

Putting It Into Practice

Install standardized AI usage as a shared operating model. Four moves the leadership team can complete inside thirty days.

THE FOUR INPUTS YOU NEED BEFORE THE STANDARDIZATION PASS

I1 — The three highest-AI-activity workflows
The three workflows the prior chapters identified, plus any departments where shadow AI usage is suspected. The discovery work focuses where activity is highest, not where leadership wants it to be highest.

I2 — The six-question conversation format
The shadow process discovery questions from this chapter. Thirty-minute conversation per workflow with one or two team members who do the work daily. The framing signals discovery, not audit.

I3 — The eight-question standard framework
Approved tools, required inputs, acceptable output, what gets checked, when human review is required, when escalation triggers, what gets documented, what is prohibited. Plain language. Single page per standard.

I4 — The five-component leadership alignment
Purpose, scope, minimum rules, metrics, exception process. The alignment conversation happens before standards reach the team, not during the rollout. Fragmented leadership signals produce inconsistency.

1 CONDUCT THE SHADOW PROCESS REVIEW

Run the six-question conversation across the three workflows. Document every finding by workflow, step, and type (better method, broken-step workaround, or unapproved risk practice). Identify the three most common undocumented habits across team members. These are the highest-priority standardization targets.

Owner: AI Efficiency Lead with Workflow Owners — within seven days

2 DRAFT THE USAGE STANDARDS

Using the eight-question framework, draft a single-page standard for each of the three highest-priority workflows. Plain language. Test each draft with one team member who runs the workflow daily before finalizing. Store the standard alongside the workflow instructions, not in a separate compliance folder.

Owner: Workflow Owner with AI Efficiency Lead

3 RUN ALIGNMENT, THEN COMMUNICATE

Hold the five-component leadership alignment conversation: purpose, scope, minimum rules, metrics, exception process. Document the agreements. Then communicate the rollout to the team with a specific non-punitive message. Acknowledge that some shadow processes were better methods that have been incorporated into the new standards.

Owner: Executive Sponsor with department heads

4 ASSIGN OWNERS AND CALENDAR THE REVIEW

Assign a named owner to each standard. The owner is responsible for spot checks, template updates, the quarterly review, and the escalation path. Put the first quarterly review on the leadership calendar before closing this action. The cadence holds only if it is scheduled.

Owner: Executive Sponsor — calendar entry within fourteen days

STANDARDIZED AI USAGE IS THE FOUNDATION ON WHICH OPTIMIZATION IS BUILT. THE STANDARDS COME FIRST. THE OPTIMIZATION FOLLOWS.

Chapter 11: The 90 Day AI Process Optimization Plan

Installing Measurable Improvement in Your AI Operating System

CHAPTER 11 OUTCOMES

At the end of this chapter, you will be able to:

1. Sequence the three thirty-day phases of the 90 Day AI Process Optimization Plan with the correct outputs at each phase boundary.
2. Select the three workflows that will be the focus of the first ninety-day cycle using a defensible prioritization rule.
3. Assign named ownership for each workflow and place the day thirty, day sixty, and day ninety reviews on the calendar before day one begins.
4. Run honest recalibration when the plan slips rather than quietly abandoning the cycle midstream.
5. Install the 90 Day Plan as a recurring operating cycle, not a one-time project. The second cycle reveals what the first cycle cannot.

EVIDENCE CREATES THE CONDITIONS FOR IMPROVEMENT. A SEQUENCED EXECUTION PLAN IS WHAT ACTUALLY PRODUCES IT.

The Problem This Chapter Solves: You Have the Frameworks. Now You Need a Sequenced Execution Plan

Ten chapters of serious operating work sit behind you at this point. You've identified the AI Efficiency Gap. You've named Phantom Productivity and measured the AI Efficiency Tax. You've built a Workflow Reality Map that shows what's actually happening inside your processes, not what the documentation says should be happening. You've run the AI Process Fit Test and assigned decisions to each active use case. You've measured cycle time, capacity, error rates, and rework. You've built the AI Efficiency Scorecard, calculated a defensible ROI number, prepared the Executive AI Efficiency Brief, and standardized AI usage across the team.

That's a complete operating foundation. Most businesses using AI don't have a fraction of it.

But here's the gap that remains. All of that work produced evidence. Evidence is not the same as execution. Knowing where the AI Efficiency Tax is highest doesn't automatically reduce it. Having a Scorecard with deteriorating use cases doesn't automatically fix them. Identifying shadow processes and writing usage standards doesn't automatically change how the team works. Evidence creates the conditions for improvement. (Mittman, 2004) A sequenced execution plan is what actually produces it.

That's what this chapter delivers. The 90 Day AI Process Optimization Plan is the operating sequence that converts everything built in the previous ten chapters into visible, measurable improvement inside the business. It's not a new layer of frameworks. It's the mechanism for running the frameworks you already have in the right order, with the right outputs, at the right pace, so that ninety days from now the leadership team can stand in front of any financial stakeholder and answer the executive question with a number, not a story.

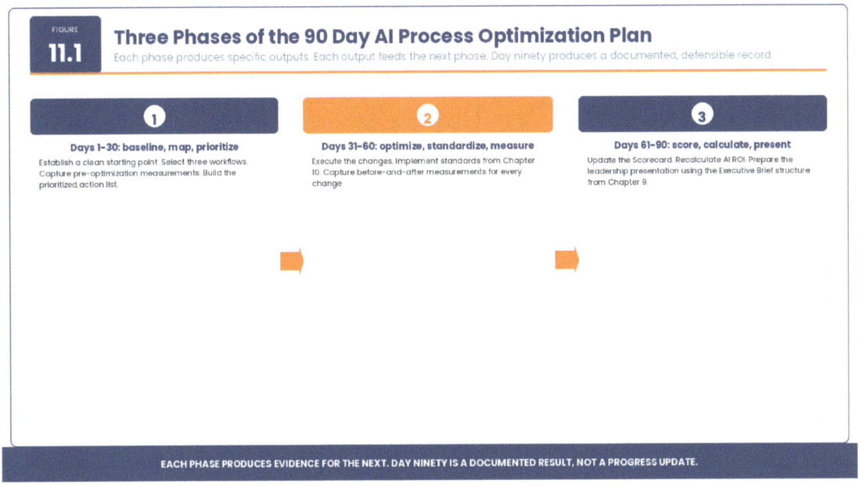

The plan runs in three phases of thirty days each. The first phase establishes a clean starting point. The second phase moves from evidence to action. The third phase converts the optimization work into documented results leadership can review and defend. Each phase produces specific outputs. Each output feeds the next phase. And at day ninety, the business has something it didn't have at day one: a documented, defensible record of what AI improved, by how much, at what cost, and what should happen next.

The promise of this chapter is specific. If the sequence is followed, the business will know what changed. It will know what it saved. It will know what it cost. And it will know exactly what to do in the next ninety-day cycle to keep improving. That's not a general productivity goal. That's an operating result with a number attached to it.

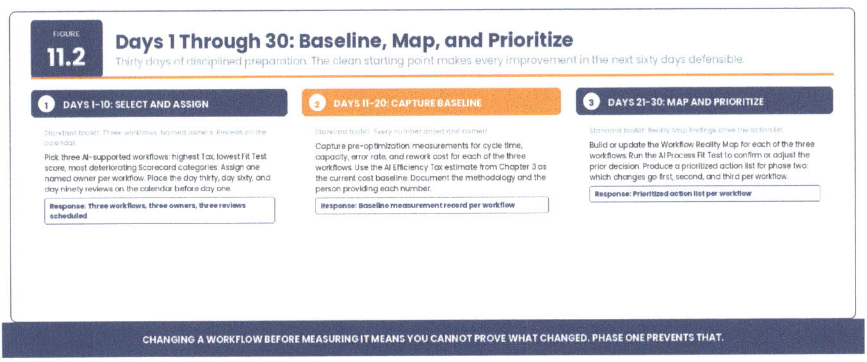

Days 1 Through 30. Baseline, Map, and Prioritize

The first thirty days are not slow. They're disciplined. That distinction matters because the instinct in most organizations is to start changing things immediately once a problem has been identified. The AI Efficiency Tax is running. The Scorecard shows deteriorating use cases. The shadow process review turned up workarounds that are costing real money. The natural impulse is to fix them right now. That impulse, when acted on without a clean starting point, almost always creates more confusion than it resolves.

Changing a workflow before you've measured its current performance means you can't prove what has changed. (Hasan, 2025) You can't compare before and after. You can't defend the improvement from the CFO who asks for the baseline. And you can't learn from the change, because you don't know what the process was actually doing before you touched it. The first thirty days prevent that problem by creating the documented starting point that makes every improvement in days thirty-one through ninety defensible.

The first task in days one through ten is selecting the three AI-supported workflows that will be the focus of this ninety-day cycle. These should be the workflows with the highest AI Efficiency Tax from Chapter 3, the lowest fit test scores from Chapter 5, or the most deteriorating Scorecard categories from Chapter 7. Pick three. Not five. Not ten. Three. The first ninety-day cycle is designed to prove the discipline on a focused set of high-impact workflows, not to fix everything simultaneously. A business that tries to optimize every AI use case in a single cycle will produce shallow improvements across many workflows rather than deep, measurable, defensible improvements (Aliyev, 2026) in the ones that matter most financially.

Once the three workflows are selected, assign a named owner to each one. The owner is responsible for completing the mapping and measurement work in phase one, executing the optimization changes in phase two, and presenting the results in phase three. Ownership is not shared across a committee. One person per workflow, with clear accountability, a defined scope, and direct access to the team members who execute the workflow daily. Without named ownership, the plan will drift. Phases will slip. Outputs will be incomplete. And the day ninety presentation will be a progress update rather than a documented result.

Days one through ten for each workflow focus on completing or updating the Workflow Reality Map. If the map was built during Chapter 4, review it now for accuracy. Workflows drift. Team practices change. (Janzen, 2025) A map that was accurate six months ago may no longer reflect how the work actually moves. Walk through the workflow with the team members doing the work and confirm that every step, every handoff, every AI touchpoint, and every informal workaround is still current. Update anything that has changed. If the map was never built, build it now using the four-step process from Chapter 4. The updated or newly built map is the first deliverable of phase one, and it's the foundation everything else in the ninety-day cycle rests on.Days eleven through twenty focus on establishing or confirming the baseline measurements for each workflow. For cycle time, measure at least fifteen consecutive cycles and calculate the average. For capacity, count the completed, usable outputs per week for the current team at the current headcount. For error rate, track first-pass accuracy over the same period. For rework, log the time spent per week on correction, cleanup, and exception handling, and multiply by loaded labor rate to produce a dollar figure. Write every number down with the date it was captured and the name of the person who confirmed it. A baseline that isn't documented with a date and a name doesn't exist for reporting purposes when day ninety arrives.

If a pre-AI baseline was captured before these tools were introduced, pull it from the records and compare it against the current state. The comparison tells the team whether the workflow has improved, stayed flat, or deteriorated since AI was introduced. That comparison is the first real finding of the ninety-day cycle, and it often changes the prioritization of what needs to be fixed. A workflow that looked like a strong candidate for optimization based on the Scorecard may turn out to have improved significantly since the last formal measurement. A workflow that looked stable may turn out to have deteriorated quietly since the last review. The baseline comparison surfaces those realities before the optimization work begins, which prevents the team from spending thirty days fixing a problem that already resolved itself or ignoring a problem that has been getting worse.

Days twenty-one through thirty focus on two specific outputs. The first is running the AI Process Fit Test for each of the three selected workflows using the current Workflow Reality Map findings as the evidence base. Score all five dimensions. Document the rationale. Assign a decision: expand, refine, pause, or remove. The fit test score tells the team what kind of optimization work is appropriate for each workflow. A workflow scoring in the refine range needs standardization and checkpoint improvements. A workflow scoring in the pause range needs process redesign before any AI optimization work begins. Knowing which category each workflow falls into before the optimization phase starts prevents the team from applying the wrong type of fix to the wrong type of problem.

The second output from days twenty-one through thirty is the prioritized action list. This is a short, ranked document that tells the business what to fix first and why. Rank the three workflows by the combination of two factors: the financial impact of the AI Efficiency Tax currently running in each workflow, and the feasibility of reducing that tax within thirty days. The workflow with the highest tax and the most achievable fix is the top priority. The workflow with the highest tax and the most complex fix is the second priority. The workflow with the lowest tax is the third. That ranking is the operating sequence for phase two, and it's the document that tells every team member and every workflow owner what to work on first when day thirty-one arrives.

By day thirty, the leadership team should have three things on paper: a current Workflow Reality Map for each selected workflow, a documented baseline for all four core metrics (introduced in Chapter 6), and a ranked action list that shows where the highest return on optimization effort is and why. Those three outputs are the clean starting point the plan requires. The business is now ready to start changing things, because it knows exactly what it's changing from, what it's changing to, and what the financial case for the change is.

Days 31 Through 60. Optimize, Standardize, and Measure

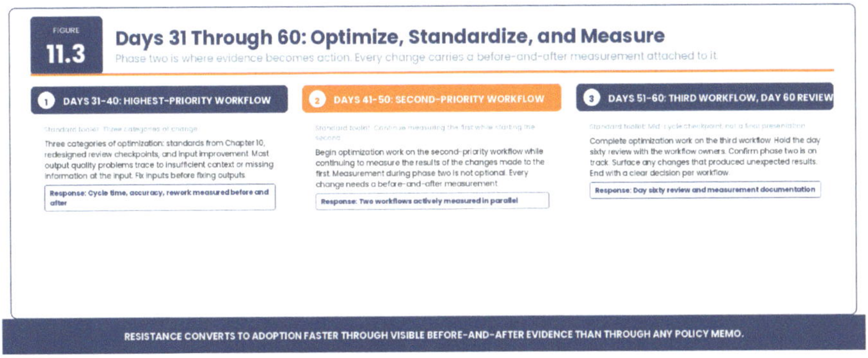

Phase two is where the evidence becomes action. The maps are built. The baselines are documented. The priorities are ranked. Now the work is to change the workflows, measure the results, and adjust quickly when a change produces an unexpected outcome. This is the phase that most organizations skip to first, and it's the phase that fails most often when it isn't preceded by the discipline of phase one. With a clean starting point, phase two is focused and fast. Without one, it's guesswork dressed up as improvement.

The first ten days of phase two, days thirty-one through forty, focus on the top-priority workflow from the ranked action list. The workflow owner leads the optimization work. The specific changes depend on what the fit test and baseline findings revealed, but the sequence for most SMB workflows follows a consistent pattern. Start with the step that generates the most review burden or the most rework. That's almost always the highest-return target, because reducing review burden and rework simultaneously recovers labor cost and improves cycle time without requiring new tools or additional headcount.

The most common optimization moves in this phase are standardization, checkpoint redesign, and input improvement. Standardization means creating or updating the shared prompting guidelines and output quality criteria for the

AI step that's producing inconsistent results. Use the eight-question framework from Chapter 10 to draft the standard. Test it with one team member before rolling it out to the full workflow. Measure first-pass accuracy before and after the standard is implemented. A well-designed prompting standard typically improves first-pass accuracy by ten to twenty percentage points within the first two weeks of consistent use, which directly reduces review time and rework cost.

Checkpoint redesign means moving the quality review to an earlier point in the workflow so that problems are caught before they compound. If the current review happens after the AI output has already been formatted, packaged, and prepared for delivery, catching an error at that stage is expensive. The error requires undoing the formatting, correcting the content, and repackaging the output. Moving the quality check to the step immediately after AI produces the raw output, before any downstream work is done with it, reduces the cost of catching errors dramatically. The output doesn't look as polished at the review stage, but the review is faster, the corrections are cheaper, and the total cycle time drops because errors don't cascade through multiple downstream steps before they're caught.

Input improvement means changing what goes into the AI tool before it runs, rather than fixing what comes out after. Most AI output quality problems trace back to insufficient context, unclear instructions, or missing information (Fischer et al., 2020) that the tool needed to produce a usable result. When the team provides more specific context, clearer parameters, and better source material before running the AI step, output quality typically improves without any change to the tool itself. The optimization work here is identifying exactly what inputs the AI step needs to produce consistent, high-quality output, documenting those inputs as required elements in the usage standard, and confirming that the team has access to those inputs before the workflow step begins.

Days forty-one through fifty focus on the second-priority workflow while continuing to measure the results of the changes made to the first. Measurement during phase two is not optional. Every change made to a workflow needs a before-and-after measurement attached to it. Cycle time before the change and cycle time after. First-pass accuracy before and after. Rework hours before and after. Without those measurements, the team can't confirm that the change produced the intended improvement, can't detect when a change created an unexpected side effect, and can't defend the optimization work to leadership at the day sixty review.

The day sixty review is a mid-cycle checkpoint, not a final presentation. Its purpose is to confirm that phase two is on track, surface any changes that produced unexpected results, and adjust the approach for the remaining ten days before phase three begins. The workflow owners present three things at the day sixty review: what was changed, what the before-and-after measurements show, and what adjustments are needed based on what the measurements revealed. The review should take no more than thirty minutes per workflow. It should end with a clear decision about whether to continue the current approach, adjust it, or escalate a problem that needs leadership attention before phase three begins.

Resistance during phase two should be expected and handled directly. (Bosse et al., 2006) Employees who have been using workarounds for months may be attached to those workarounds, not because they're being difficult, but because the workarounds solved a real problem and the new standard hasn't yet proven it can solve the same problem better. The response to that resistance is not enforcement. It's evidence. Show the team member the before-and-after measurements for the change. Show them the rework reduction. Show them the cycle time improvement. When the evidence is visible and specific, resistance usually converts to adoption faster than any policy memo or management directive (Luecke, 2003) ever could.

Days fifty-one through sixty focus on the third-priority workflow and on completing the measurement documentation for all three workflows before phase three begins. Every change made in phase two needs a documented measurement record: what the baseline was, what changed, when the change was made, and what the post-change measurement shows. That documentation is the evidence base for the day ninety presentation, and it needs to be complete and organized before phase three starts. A team that spends the first week of phase three scrambling to reconstruct measurement records will produce a weaker presentation than a team that enters phase three with every number already documented and ready to report.

The key output of phase two is a set of measurable improvements with documented before-and-after data for each of the three selected workflows. Not estimates. Not impressions. Documented measurements that show what changed, by how much, and what it cost to produce the change. That's the evidence that phase three converts into financial results and leadership presentations.

Days 61 Through 90. Score, Calculate, and Present

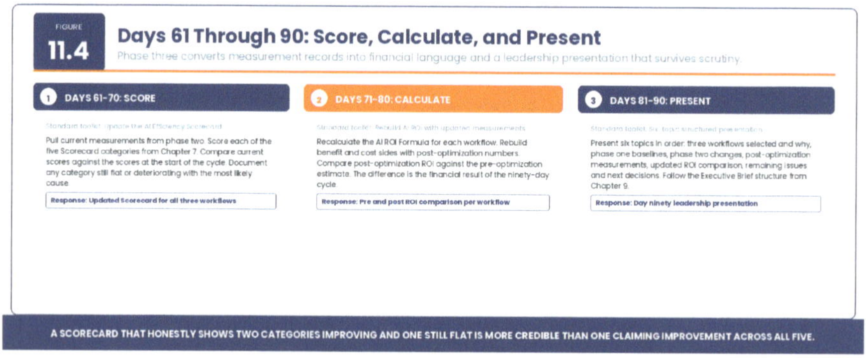

Phase three is where the optimization work becomes a business result. The changes have been made. The measurements are documented. Now the task is to translate those measurements into financial language, update the operating tools that track AI performance, and prepare a leadership presentation that shows the business exactly what the ninety-day cycle produced and what it should do next.

Days sixty-one through seventy focus on updating the AI Efficiency Scorecard for each of the three workflows. Pull the current measurements from the phase two documentation. Score each of the five categories using the criteria from Chapter 7: cycle time change, capacity impact, error rate trend, rework frequency, and estimated financial impact. Compare the current scores against the scores from the beginning of the ninety-day cycle. The comparison tells the story of what the optimization work accomplished in operating terms, and it's the first layer of the evidence that the day ninety presentation will be built on.

If any category is still scoring flat or deteriorating after the phase two changes, document the finding specifically. What was done, what the measurement shows, and what the most likely cause of the remaining gap is. Don't overstate the improvement by ignoring the categories that didn't move. A Scorecard that honestly shows two categories improving and one category still flat is more credible than a Scorecard that claims improvement across all five when the measurements don't support it. The credibility of the ninety-day cycle depends entirely on presenting what the data actually supports, not what leadership hoped the data would show.

Days seventy-one through eighty focus on recalculating the AI ROI Formula for each workflow using the updated measurements. Rebuild the benefit side with the post-optimization numbers. Cycle time savings after the changes. Capacity gains that were converted into financial value. Rework cost eliminated as a result of the standardization and checkpoint redesign work. AI Efficiency Tax reduction compared to the baseline from the beginning of the cycle. Add those four components to get the updated annual benefit figure. Keep the cost denominator complete. Include the subscription cost, the implementation time for the optimization changes made in phase two, the training time for any new standards introduced, and the ongoing review burden at its new, reduced level. Divide the updated benefit by the updated cost. That's the post-optimization ROI for each workflow.

Compare the post-optimization ROI against the pre-optimization estimate from Chapter 8. The difference between those two numbers is the financial result of the ninety-day cycle. If the pre-optimization ROI was negative fourteen percent and the post-optimization ROI is positive twenty-two percent, the ninety-day cycle produced a thirty-six percentage point improvement in return for that use case. That's a specific, defensible financial result. It has a methodology behind it. It has documented measurements supporting it. And it can be presented to any financial stakeholder without requiring a thirty-minute explanation of how it was calculated.

Also confirm whether the AI Efficiency Tax decreased for each workflow. The tax estimate from Chapter 3 is the before number. The updated tax calculation, using the post-optimization review burden, rework cost, and management overhead measurements, is the after number. The difference is the monthly tax reduction. Multiply by twelve for an annual figure. That number represents the margin that the ninety-day cycle recovered, and it belongs in the financial summary of the day ninety presentation alongside the ROI comparison.

Days eighty-one through ninety are dedicated to preparing and delivering the leadership presentation. The presentation format follows the Executive AI Efficiency Brief structure from Chapter 9, with one addition specific to the ninety-day cycle: a before-and-after comparison section that shows the baseline at day one, the changes made in phase two, and the measured results at day ninety. That comparison is the core of the presentation. Everything else supports it.

The presentation should cover six specific topics in order. First, the three workflows that were selected and why they were prioritized. Second, the baseline measurements captured in phase one, with the documented date and source for each number. Third, the specific changes made in phase two, including what was standardized, what checkpoints were redesigned, and what input improvements were implemented. Fourth, the post-optimization measurements for each workflow, compared directly against the baseline. Fifth, the updated ROI calculation for each workflow, showing the pre-optimization and post-optimization figures side by side with the financial impact of the difference. Sixth, the remaining issues and the next recommended decisions, meaning which workflows should be expanded, which need another optimization cycle, and which new workflows should be selected for the next ninety-day plan.

The remaining issues section is as important as the results section. A presentation that shows only improvements and ignores the gaps that remain will be viewed skeptically by any financially sophisticated stakeholder who knows that no ninety-day cycle fixes everything. A presentation that honestly identifies what improved, what didn't, and what needs to happen next demonstrates that leadership is managing AI with discipline rather than reporting selectively. The remaining issues section is also where the next cycle's priorities come from, which connects the day ninety presentation directly to the day one work of the next cycle and keeps the operating discipline running continuously rather than stopping after the first success.

One specific warning for the day ninety presentation. Don't present a number you can't defend under a follow-up question. If a CFO asks how the rework reduction was calculated, the answer needs to be immediate and specific: the team tracked rework hours per week for thirty days before the optimization changes and thirty days after, multiplied the difference by the loaded labor rate, and here's the documentation. If the answer to that question is "we estimated it based on what the team told us," the number will be challenged and the credibility of the entire presentation will be reduced. Build the documentation before the presentation, not after someone asks for it.

Installing the 90 Day Plan as a Recurring Operating Cycle

The ninety-day cycle shouldn't end as a completed project. It should become the quarterly rhythm for improving AI performance inside the AI Operating System. The first cycle proves the method. The recurring cycle makes the improvement durable. That's the distinction that separates a business that ran one successful AI optimization project from a business that runs a continuous AI operating discipline that gets stronger every quarter.

The transition from a first cycle to a recurring cadence requires three specific structural changes. Each one is simple. None of them are optional if the recurring cycle is going to hold.

The first structural change is ownership distribution. The first ninety-day cycle is often driven by one person, a COO, an operations director, or a designated AI champion who pushed the process through from day one to day ninety. That's appropriate for a first cycle. It's a problem for a recurring one. A recurring cycle that depends on one person to initiate, coordinate, and present every quarter will fail the moment that person is unavailable, changes roles, or simply gets pulled into other priorities. The recurring cycle needs to be owned by the workflow owners and department heads who are responsible for the workflows being optimized. Those people need to own the measurement updates, the Scorecard reviews, the fit test re-scores, and the optimization work between cycles. The central champion's role shifts from doing the work to facilitating the process and holding the owners accountable for their deliverables.

To make that ownership shift practical, assign each active AI use case a named workflow owner before the second cycle begins. That owner is responsible for four recurring tasks: maintaining the Workflow Reality Map for their workflow, updating the AI Efficiency Scorecard at the end of each quarter, completing the AI Process Fit Test re-score, and presenting the optimization priorities and results at the quarterly leadership review. Those four tasks are specific, time-bounded, and achievable by a department head or team lead without requiring dedicated AI expertise. They're the minimum recurring responsibilities that keep the AI operating system running without a central champion doing all of the work.

The second structural change is calendar integration. AI optimization must become part of how the business runs, not a special initiative that runs alongside the normal operating rhythm. When the ninety-day cycle lives on a separate project calendar, it competes with every other priority in the business for leadership attention. When it's built into the same operating calendar as the quarterly financial review, the monthly operations meeting, and the annual planning process, it becomes part of the standard rhythm rather than an additional demand on leadership time.

The specific calendar integration looks like this. At the start of each quarter, the workflow owners complete the Workflow Reality Map updates and the fit test re-scores for their assigned workflows. At the midpoint of the quarter, the day sixty review happens as a standing agenda item in the regular operations meeting. At the end of the quarter, the AI Efficiency Scorecard is updated, the ROI calculations are refreshed, and the Executive AI Efficiency Brief is prepared for the leadership review. The leadership review happens as part of the existing quarterly business review, not as a separate AI meeting. That integration is what keeps the discipline running without requiring a new organizational structure or a new meeting cadence.

The third structural change is connecting the ninety-day cycle outputs to the business's financial planning process. The ROI calculations from each cycle should feed directly into the AI budget conversation for the following quarter. Use cases with strong, improving ROI are candidates for expanded investment. Use cases with flat or declining ROI are candidates for reduced investment or reallocation. Use cases with negative ROI that haven't improved after two consecutive cycles are candidates for removal. When the ninety-day cycle outputs inform budget decisions, AI optimization becomes financially self-reinforcing. Strong performers get more resources. Weak performers get fixed or removed. And the overall AI investment concentrates in the use cases that are producing measurable returns rather than spreading evenly across all use cases regardless of performance.

The strongest message about the recurring cycle is this. The first ninety days prove that the method works. They show the leadership team that structured workflow mapping, baseline measurement, focused optimization, and disciplined reporting can produce a defensible financial result in a short time frame without requiring a major operational overhaul. That proof is valuable. But it's not the goal. The goal is a business where AI performance improves every quarter because the operating system for measuring, optimizing, and reporting AI performance is running continuously, owned by the right people, integrated into the normal business rhythm, and connected to the financial decisions that determine where AI investment goes next.

Sequencing the First Three Workflows for the 90 Day Plan

The 90 Day Plan recommends selecting three workflows for the first cycle. The number three is not arbitrary. One workflow does not generate enough operating learning to inform the next cycle. Five workflows split leadership attention too thin and produce shallow results across all of them. Three is the operating sweet spot for SMB AI optimization: enough to produce a portfolio of learnings, few enough to allow real attention to each one. The question that determines whether the first 90 days produces a strong second cycle or a weak one is how the three workflows are selected.

The selection pattern that works best in SMB engagements is the broad-plus-deep combination. One of the three workflows is selected because it is the highest-impact AI use case in the business, regardless of how much optimization attention it has received. This workflow gets the deep treatment: full Reality Map, complete fit test, comprehensive measurement, and a focused improvement plan. The second workflow is selected because it represents a different operating function from the first, broadening the leadership team's understanding of how AI behaves across functions. The third workflow is selected because it is a candidate for removal or pause, and the ninety-day cycle is the opportunity to make the formal decision based on measurement rather than impression.

The broad-plus-deep pattern produces three different kinds of learning. The deep workflow teaches the team what a fully optimized AI use case looks like inside their business, which becomes the operating reference for subsequent cycles. The broad workflow teaches the team that the patterns from the deep workflow do not always transfer cleanly to other functions, which prevents over-generalization of the lessons. The pause candidate teaches the team how to use the Volume IV instruments to make a hard decision and document it defensibly, which is the muscle the team will need for subsequent cycles when other workflows reach the same decision point.

The alternative selection pattern (three workflows from the same function) is less informative. Optimizing three sales workflows in the first cycle teaches the team about sales AI optimization, which is useful, but does not build the cross-functional pattern recognition that the broad-plus-deep approach produces. Some businesses choose the same-function pattern for valid reasons (the AI program is concentrated in one function, or the leadership team wants depth in one area before going broader), and the pattern works in those contexts. But for most SMB AI programs, the broad-plus-deep approach produces a stronger second cycle and a more capable leadership team going forward.

The sequencing within the 90 days also matters. The deep workflow should be the focus of the first thirty days, because the foundational work (Reality Map, baseline, fit test) sets the pattern that the other two workflows will follow. The broad workflow comes into focus during days thirty-one through sixty, by which point the team has internalized the methodology from the deep workflow and can apply it more efficiently to the second case. The pause candidate is sequenced into the last thirty days, when the team has enough comparative experience to make a defensible decision about whether the workflow belongs in the next cycle. The sequencing distributes the leadership team's attention across the 90 days rather than concentrating it on day one.

When the Plan Slips: Honest Recalibration vs Quiet Abandonment

FIGURE 11.5

When the Plan Slips: Honest Recalibration vs Quiet Abandonment

Every ninety-day cycle hits at least one moment where something is not on schedule. The response defines whether the cycle survives.

HONEST RECALIBRATION
What recovery actually looks like

Name the slip explicitly
Surface it at the next review. State which workflow, which output, which date was missed, and why.

Recalibrate the timeline
Adjust the dates with documented reasoning. The cycle still ends at day ninety. The internal milestones shift to reflect reality.

Document the cause
Capture what slowed the work. The cause becomes input to the next cycle's planning, so the slip is not repeated.

Continue measurement
Every change still carries before-and-after data. The cycle slows but it does not lose its evidence base.

QUIET ABANDONMENT
The pattern to avoid

Reviews get postponed
The day sixty review slips by two weeks. Then four. Then it does not happen. The cycle loses its checkpoints.

Workflows get dropped
The hardest workflow gets quietly removed from the cycle. The leadership team stops mentioning it.

Documentation stops
Before-and-after measurements stop being captured. The evidence base for day ninety dissolves.

Day 90 presentation becomes vague
What was supposed to be a documented result becomes a progress update. Credibility costs more than the original slip would have.

The operating discipline: A slip is information. Treat it as data the next cycle needs, not as a failure to be hidden. The leadership team that recalibrates honestly produces stronger second and third cycles than the team that protects the first cycle's appearance by quietly abandoning it.

RECALIBRATION IS RECOVERY. ABANDONMENT IS A CREDIBILITY LOSS THAT COMPOUNDS ACROSS CYCLES.

90 Day Plans slip. The first cycle of any new operating discipline takes longer than the planned duration, runs into unexpected complications (Soofastaei, 2025) , and reaches the ninety-day checkpoint without all three workflows fully through the cycle. This is not a failure of the plan or the team. It is the normal pattern for the first cycle. What matters is how the leadership team responds to the slip. The response determines whether the second cycle starts in a stronger position than the first or whether the entire plan quietly fades.

The diagnostic moment is the day-thirty check-in. The plan calls for the Reality Maps to be complete, the baselines documented, the fit tests scored, and the optimization priorities ranked by day thirty. If these are complete, the plan is on track. If they are partially complete, the plan is slipping. The honest response to a partial day-thirty position is to recalibrate the remaining 60 days based on what is realistic given the current pace, not to maintain the original plan against evidence that the pace will not support it. The recalibrated plan covers fewer optimization actions in the remaining time but completes them properly. The unmodified plan attempts everything and completes nothing.

The day-sixty check-in is the second diagnostic moment. By day sixty, the plan calls for standardization improvements, checkpoint redesigns, and before-and-after measurements to be in progress for the priority workflows. If the work is progressing, the plan is recovering. If the work has stalled, the cause needs to be identified. The most common causes are leadership attention diverted to other operating issues, team capacity consumed by the day-to-day work, or unanticipated complexity in one of the workflows that has absorbed disproportionate effort. Each cause has a different corrective response. The diagnosis is what determines which corrective response applies.

The quiet abandonment pattern is the alternative to honest recalibration. The plan was launched with enthusiasm. The first few weeks produced real work. Then competing priorities began to absorb leadership attention. The day-thirty check-in was deferred. The day-sixty check-in was deferred. Day ninety arrived without anyone formally noting that the plan had not been executed. The team is no worse off than they were before the plan started, but the operating credibility of the leadership team has eroded slightly. The next time leadership proposes a 90 Day Plan, the team will respond with more skepticism than they did the first time.

The discipline that prevents quiet abandonment is the calendar discipline. The day-thirty, day-sixty, and day-ninety review dates are scheduled on the leadership calendar before day one begins, with named owners and required deliverables. If a review date is approaching and the deliverables are not in place, the response is to acknowledge the position honestly, decide whether to recalibrate or to extend, and document the decision. The acknowledgment in writing is what separates honest recalibration from quiet abandonment. A leadership team that consistently acknowledges position relative to plan, even when the position is behind schedule, maintains the operating credibility (Kerzner, 2025) that the planning discipline depends on.

What the Second Cycle Reveals That the First Cannot

The first 90 Day Plan produces real results, but its most important output may be the conditions for the second cycle. The second cycle benefits from learnings that did not exist before the first cycle ran. The team has operated the Reality Map process, the fit test process, the Scorecard, the ROI calculation, the standardization work, and the Brief production. The second cycle uses these instruments more fluently, produces stronger results in less time, and surfaces the patterns the first cycle could not see.

The most important second-cycle pattern is the workflow-to-workflow learning transfer. In the first cycle, each workflow is optimized largely in isolation, because the team is learning the methodology while applying it. In the second cycle, the team can apply learnings from the first cycle's workflows to the second cycle's workflows. A standardization technique that worked on one workflow can be tried on another. A measurement methodology that produced clean data in one case can be transferred. A fit test scoring approach that aligned the team can be reused. The transfer reduces the time required for each subsequent workflow's optimization, and the team's effective capacity for optimization grows over each cycle.

The second cycle also reveals the durability of the first cycle's improvements. Workflows that were optimized in the first cycle either maintain their gains, drift back toward the pre-optimization state, or continue to improve as the team operates them. Each pattern carries different implications. Workflows that maintain their gains validate the standardization work and signal that the discipline is taking hold. Workflows that drift back signal that the standardization needs reinforcement or that underlying conditions have changed. Workflows that continue to improve signal that the optimization unlocked a path the team is still walking down. The second cycle's Scorecard data, read against the first cycle's, surfaces these patterns clearly.

The third pattern that the second cycle reveals is the leadership team's operating cadence around the AI program. The first cycle was something new the team was doing. The second cycle is part of the operating rhythm. The reviews happen on schedule, the deliverables are produced on time, the Brief is presented in the expected operating meeting. By the third cycle, this rhythm becomes second nature, and the AI program operates the way the rest of the business operates: with regular cadences, named owners, documented decisions, and continuous measurement. The third cycle is when the Volume IV work has fully installed, and the operating discipline becomes durable in a way that survives leadership turnover and team changes.

A business that reaches that state has done something genuinely rare. It has moved from AI adoption to AI operating discipline. It can answer the executive question not just once, at the end of a ninety-day project, but every quarter, with updated numbers, documented improvements, and a clear view of what comes next. That's what the AI Operating System is designed to produce. And the 90 Day AI Process Optimization Plan, run as a recurring quarterly cadence, is the mechanism that keeps it running.

Putting It Into Practice

The 90 Day AI Process Optimization Plan converts the evidence built across the previous ten chapters into visible, measurable improvement inside the business. It runs in three phases. Phase one, days one through thirty, creates the clean starting point: updated Workflow Reality Maps, documented baselines for all four core metrics, completed AI Process Fit Test scores, and a prioritized action list ranked by financial impact and feasibility. Phase two, days thirty-one through sixty, moves from evidence to action: standardization improvements, checkpoint redesigns, input quality upgrades, and before-and-after measurements for every change made. Phase three, days sixty-one through ninety, converts the optimization work into documented results: updated Scorecard scores, recalculated ROI figures, a confirmed AI Efficiency Tax reduction, and a leadership presentation that shows the baseline, the changes, the measured results, the financial impact, the remaining issues, and the next recommended decisions.

Each phase produces a specific output. Phase one produces the prioritized action list. Phase two produces the before-and-after measurement record. Phase three produces the day ninety leadership presentation. Those three outputs, taken together, give the business a documented, defensible record of what the ninety-day cycle accomplished and what it should do next. That record is what makes the recurring cycle possible, because each cycle's day ninety output becomes the next cycle's day one starting point.

The recurring cycle is what makes the improvement durable. Ownership shifts from a single champion to the workflow owners and department heads. The cycle integrates into the existing operating calendar. The outputs connect to the financial planning process. And every quarter, the business gets better at measuring AI performance, reducing the AI Efficiency Tax, and concentrating investment in the use cases that are producing measurable returns.

Now complete these specific steps to launch the first cycle.

First, select the three AI-supported workflows that will be the focus of this first ninety-day cycle. Use the AI Efficiency Tax estimates from Chapter 3, the fit test scores from Chapter 5, and the Scorecard findings from Chapter 7 to make the selection. Write the three workflows down by name. Confirm that each one has a named owner before moving to the next step.

Second, schedule the three key review dates on the leadership calendar before day one of the cycle begins. The day thirty review is the phase one output review: Workflow Reality Maps, baselines, and the prioritized action list. The day sixty review is the mid-cycle checkpoint: changes made, before-and-after measurements, and adjustments needed. The day ninety review is the final presentation: updated Scorecard, recalculated ROI, tax reduction confirmation, and next cycle recommendations. Put all three dates on the calendar now. A cycle without scheduled review dates will drift. Scheduled reviews create the accountability that keeps the phases on track.

Third, brief the three workflow owners on their phase one responsibilities before day one. Each owner needs to know exactly what they're responsible for producing by day thirty: an updated Workflow Reality Map, documented baseline measurements for all four core metrics, a completed AI Process Fit Test score, and their workflow's ranking on the prioritized action list. Give each owner the specific tools from the previous chapters they'll need to complete that work. Don't assume they remember the details from earlier in the book. Walk through the relevant frameworks with them directly so that phase one starts with clarity rather than confusion about what "done" looks like.

Fourth, communicate the launch to the broader leadership team with a specific, direct message. The business is beginning its first ninety-day AI optimization cycle. Three workflows have been selected. Three owners have been assigned. Three review dates are on the calendar. The goal of the cycle is to produce documented, defensible improvements in cycle time, capacity, error rates, rework, and ROI for the selected workflows, and to present those results to the leadership team at day ninety. That message sets the expectation that this is a structured operating initiative with specific outputs and specific accountability, not a general improvement effort that will be reviewed "when we have time."

Do not begin the plan with every AI use case in the business. The first cycle should prove the discipline on three focused, high-impact workflows. A focused first cycle that produces two or three strong, documented results is worth more than a broad first cycle that touches ten workflows and produces shallow, undocumented improvements across all of them. The first cycle is the proof of concept. The recurring cadence is where the scope expands. Prove the method first. Scale it second.

The business that completes this first cycle will have something it didn't have ninety days earlier. Not just better AI performance in three workflows. A proven operating sequence for improving AI performance in any workflow, documented evidence that the sequence produces measurable financial results, a leadership team that has practiced the discipline and knows how to run it, and a starting point for the next cycle that is stronger, more precise, and more financially grounded than the one that launched the first. That compounding effect, cycle over cycle, quarter over quarter, is what the AI Operating System is built to produce. The first ninety days start it. Every cycle after that builds on it.

FIGURE 11.6

Putting It Into Practice

Launch the first 90 Day cycle on three focused workflows. Three owners. Three review dates. Documented results at day ninety.

THE FOUR INPUTS YOU NEED BEFORE DAY ONE

I1 — The three workflows for the first cycle
Three workflows selected using the Chapter 3 Tax estimates, the Chapter 5 Fit Test scores, and the Chapter 7 Scorecard findings. Three, not ten. A focused first cycle proves the method.

I2 — Three named workflow owners
One named owner per workflow before day one. Confirmed in writing. An owner without a name is not an owner; it is an expectation that the work will somehow happen.

I3 — Three review dates on the calendar
Day 30 (phase one output review), day 60 (mid-cycle checkpoint), day 90 (final presentation). All three on the leadership calendar before day one. A cycle without scheduled reviews drifts.

I4 — The phase one toolkit
The Reality Map template from Chapter 4, the Fit Test from Chapter 5, the four indicators framework from Chapter 6, the Scorecard structure from Chapter 7. The owners need the tools, not summaries of them.

1 SELECT THE THREE WORKFLOWS

Use the Tax estimates, Fit Test scores, and Scorecard findings from prior chapters to make the selection. Write the three workflows down by name. Confirm each one has a named owner before moving to the next step. The first cycle proves the method on three focused workflows, not the entire AI portfolio.

Owner: Executive Sponsor with AI Efficiency Lead

2 SCHEDULE THE THREE REVIEW DATES

Day 30: Reality Maps, baselines, and the prioritized action list reviewed. Day 60: mid-cycle changes, before-and-after measurements, and any course corrections. Day 90: final presentation with updated Scorecard, recalculated ROI, Tax reduction, and next-cycle recommendations. All three dates on the calendar before day one.

Owner: Executive Sponsor — calendar entries before day one

3 BRIEF THE WORKFLOW OWNERS

Each owner knows exactly what they are responsible for producing by day 30: an updated Reality Map, baseline measurements for the four indicators, a completed Fit Test score, and their workflow's position on the prioritized action list. Walk through the relevant frameworks with each owner directly. Phase one starts with clarity, not with confusion about what "done" looks like.

Owner: AI Efficiency Lead with each Workflow Owner

4 COMMUNICATE THE LAUNCH

Send a direct message to the leadership team: the business is beginning its first ninety-day cycle, three workflows have been selected, three owners are assigned, three review dates are scheduled, and the goal is documented improvements in cycle time, capacity, error rates, rework, and ROI. The message frames this as a structured operating initiative with specific outputs and accountability.

Owner: Executive Sponsor — sent before day one

PROVE THE METHOD FIRST. SCALE IT SECOND. A FOCUSED FIRST CYCLE PRODUCES THE EVIDENCE THE NEXT ONE BUILDS ON.

Chapter 12: AI Efficiency and Business Valuation

Making Your AI Operating System a Financial Asset

CHAPTER 12 OUTCOMES

At the end of this chapter, you will be able to:

1. Recognize that AI operating discipline has become a financial due diligence category and affects business valuation directly.
2. Assemble the Due Diligence AI Operations Package with an executive summary layer and a ten-document supporting layer.
3. Tailor the package for acquirers, lenders, and investors by emphasizing the documentation they each weigh most.
4. Operate the eleven-step action sequence that compounds across quarters into a defensible AI Operating System.
5. Apply the twenty operating statements (ten to do, ten not to do) as the practical capstone of Pillar I.

AI VALUE IS NOT ONLY CREATED BY USING THE TOOLS. IT IS CREATED BY PROVING THE BUSINESS CAN MANAGE, MEASURE, AND DEFEND THEIR USE.

Everything built across the previous eleven chapters was always heading somewhere specific. The Workflow Reality Maps, the AI Process Fit Test scores, the AI Efficiency Scorecard, the ROI calculations, the Executive AI Efficiency Brief, the 90 Day AI Process Optimization Plan. Each one was valuable on its own. Together, they form something more significant than a collection of operating tools. They form a documented, measurable, defensible record of how AI is being managed inside the business. And that record has a financial consequence that most leaders haven't fully considered yet.

It affects how the business is valued.

Not in a theoretical way. In a direct, practical way that shows up in acquisition conversations, lender reviews, investor updates, and due diligence processes. The business that can walk into any of those conversations with a complete AI operating record, including documented workflows, measured efficiency gains, calculated ROI, and a structured quarterly improvement cadence, looks fundamentally different from the business that can only say its team is using AI tools and adoption is high. One looks managed. The other looks uncertain. And financial stakeholders, without exception, pay more for certainty and discount uncertainty (Buckley et al., 2021) .

This Volume closes the loop.

The Problem This Chapter Solves: Most Leaders Don't Realize Their AI Discipline Affects How the Business Is Valued

Most leaders think about AI performance as an internal management question. Did the workflow get faster? Did rework go down? Is the team using the tools consistently? Those are the right questions to ask inside the operation. But they're not the only questions that matter. There's a second set of questions that financial stakeholders ask, and those questions have a direct impact on how the business is perceived, how it's priced, and how much confidence a buyer, lender, or investor is willing to place in its future performance.

Those questions sound like this. How is AI being used in this business? What has it improved, and can you show us the numbers? What risks does AI introduce, and what controls are in place to manage them? Who owns AI quality and performance? What documentation supports your answers to any of these questions? And if we asked to see your AI operating record tomorrow, what would you be able to produce?

Most leadership teams aren't ready for those questions. Not because they've done bad work, but because they've been managing AI as an operational tool without recognizing that AI operations have become a financial due diligence category. That shift has happened quietly over the past two years, and it's accelerating. Buyers, lenders, and investors are increasingly sophisticated about AI, and they're no longer satisfied with enthusiasm and adoption statistics. They want to see operating evidence. They want documentation. They want a measurement system. They want to know that AI inside the business is governed, measured, and improving rather than scattered, undocumented, and generating hidden cost.

A business with scattered AI usage, weak measurement, undocumented workflows, and no clear ownership doesn't look innovative to a financially sophisticated stakeholder. It looks risky. Because undocumented AI activity creates uncertainty around margin, labor productivity, data handling, scalability, and management control. And uncertainty, in financial conversations, translates directly into risk adjustments, lower valuations, higher interest rates, more demanding covenants (Demerjian, 2017) , or a slower path to closing a transaction.

The AI Operating System built across this book addresses every one of those concerns. But only if leadership understands the connection between operating discipline and financial perception, and only if that discipline is documented in a way that can be placed in front of a financial stakeholder and held up under scrutiny. That's what this chapter is about. Not adding new frameworks. Showing how the frameworks already built become a financial asset when they're assembled into a complete, defensible operating record.

The key idea is straightforward. AI value isn't only created by using the tools. It's created by proving that the business can manage, measure, improve, and defend their use over time. That proof is what separates a business that talks about AI from a business that has built AI into its operating discipline. And that distinction has a financial consequence that shows up in every high-stakes conversation the business will ever have with a financial stakeholder.

How AI Efficiency Documentation Affects Acquisition, Lending, and Investment Conversations

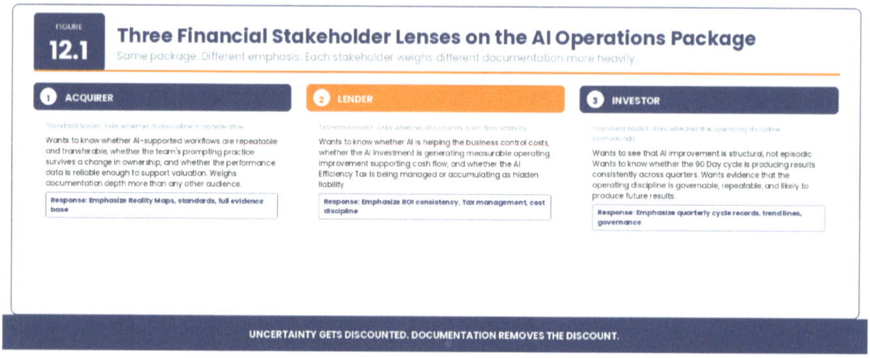

Financial stakeholders approach AI from a risk management perspective first and an opportunity perspective second (Giudici, 2018) . That ordering matters because it determines what they're looking for when they ask about AI operations. They're not primarily asking whether AI is exciting or whether the team is enthusiastic. They're asking whether AI makes the business more predictable, more scalable, more defensible, and more valuable. Or whether it introduces hidden cost, operational uncertainty, data exposure, and management complexity that makes the business harder to underwrite, harder to acquire, and harder to trust.

Buyers evaluating an acquisition target want to know whether AI-supported workflows are repeatable and transferable. A workflow that depends entirely on one employee's personal prompt library, runs through an unapproved tool, and has never been formally documented or measured isn't a business asset. It's a personal practice that may or may not survive a change in ownership. A workflow that has a documented Workflow Reality Map, a usage standard, a named owner, a baseline measurement, and a current efficiency score is a documented operating process that transfers with the business. That distinction changes how an acquirer thinks about the reliability of the revenue and margin the workflow supports.

The consequence of missing documentation is not abstract. A 75-person construction firm entered an acquisition conversation eighteen months after introducing AI across its invoice processing and project reporting workflows. The buyer's diligence team asked four specific questions about AI operations: which tools, what they cost, what they produced, and how performance was being managed. The firm had usage reports from the vendor. It did not have a Scorecard, a Tax estimate, or a documented ROI. The diligence findings flagged AI as an undocumented operational variable. The acquisition closed at a price that reflected the variable as risk. The buyer's letter of intent had referenced a multiple range; the final number landed at the lower bound. The firm's leadership team learned, after the close, that an AI operations package built across two quarters would likely have changed the conversation. The cost of not building it was not theoretical. It was a measurable difference in the sale price, paid in full at closing.

Lenders evaluating creditworthiness want to know whether AI is contributing to operating stability or introducing volatility. A business that can show consistent, documented efficiency improvements in its AI-supported workflows is demonstrating that its operating costs are under control and its productivity is improving. A business that can't explain what AI is doing to its labor model, its error rates, or its capacity is introducing a variable that a lender can't model. Lenders don't like variables they can't model. They price them as risk (Strahan, 1999) , which means higher rates, tighter covenants, or more conservative loan structures.

Investors and board members want to know whether AI is a documented source of margin improvement or an untracked cost center. The AI Efficiency Scorecard, the ROI Formula calculations, and the Executive AI Efficiency Brief answer that question directly. They show whether AI use cases are producing positive returns, what those returns are in dollar terms, and what the business is doing to expand the ones that are working and fix the ones that aren't. That's the kind of structured, evidence-based AI narrative that earns investor confidence. It's also the kind that holds up when an investor asks a follow-up question, because every number in the narrative traces back to documented measurement work.

The frameworks built across this book answer financial stakeholders' questions before those questions are asked. The AI Tool Inventory shows what's being used and what it costs. The Workflow Reality Maps show how AI is embedded in real operating processes. The AI Process Fit Test scores show whether AI is placed correctly or generating risk. The AI Efficiency Scorecard shows whether performance is improving or deteriorating. The ROI Formula calculations show the financial return after all costs are included. The Executive AI Efficiency Brief packages all of that into a format any financial stakeholder can read and trust. And the 90 Day AI Process Optimization Plan shows that improvement is a structured, recurring discipline rather than a one-time project.

One important clarification deserves to be stated directly. Strong AI documentation doesn't guarantee a higher valuation itself. Valuation is driven by many factors (Moro-Visconti, 2024) , and AI operating discipline is one input among several. What documentation does is reduce uncertainty. It removes the discount that financial stakeholders apply when they can't see clearly (Jennings, 2024) what AI is doing inside the business. It replaces vague claims with specific evidence. And it allows the business to make a credible case that AI is contributing to operating performance rather than quietly consuming margin. That case, made with documented evidence rather than enthusiasm, strengthens the business's position in every financial conversation it enters. Financial stakeholders trust evidence more than they trust stories (Igwe-Nmaju & Anadozie, 2022) . This book gives leadership the evidence package they need to tell a story that holds up under scrutiny.

FIGURE 12.2 How Operating Discipline Becomes a Financial Asset
Each instrument from this Volume contributes to the documented evidence package that affects valuation.

Building an AI Operating System That Gets Stronger Every Quarter

The businesses that gain lasting value from AI won't be the ones that adopted the most tools the fastest. They'll be the ones that turned AI into a disciplined operating system that compounds in value every quarter it runs. That's a specific kind of advantage, and it's built through a specific kind of operating discipline. Not a single impressive AI deployment. A recurring cycle of measurement, improvement, documentation, and reporting that makes the AI operation more efficient, more reliable, and more financially defensible with every pass through the cycle.

The frameworks in this book are designed to work as a compounding system. Each one feeds the next, and the outputs of one cycle become the starting point for the next. The Workflow Reality Map shows what's actually happening inside the process. The AI Process Fit Test determines where AI belongs and where it doesn't. The efficiency metrics from Chapter 6 prove what moved. The AI Efficiency Scorecard consolidates the results into a single operating view. The AI ROI Formula converts improvement into financial value. The 90 Day AI Process Optimization Plan keeps the improvement cycle moving. And the Executive AI Efficiency Brief makes the results visible to the financial stakeholders who need to see them.

Each quarter that cycle runs, something specific happens. Documentation gets cleaner. Workflows get more standardized. Output quality improves as prompting guidelines mature and team members gain experience with the standards. Review burden declines as first-pass accuracy rises. Rework costs drop as the AI Efficiency Tax is systematically reduced. ROI figures improve as the benefit side of the formula grows and the cost side shrinks. And the overall AI operating record becomes thicker, more precise, and more defensible with each completed cycle.

That compounding effect has a financial consequence that goes beyond the operational improvements themselves. A business with one quarter of documented AI optimization looks like it's trying to get organized. A business with four consecutive quarters of documented AI optimization, each one showing measurable improvement over the previous, looks like it has an operating discipline that's working. The difference between those two impressions is significant in a financial due diligence context, because the second business isn't just showing results. It shows a system that produces results reliably and repeatedly. That's what scalability looks like from a financial stakeholder's perspective, and scalability commands a premium (Moro-Visconti, 2024).

Consider a hypothetical scenario that makes this concrete. Think about two businesses in the same industry, both using AI across similar workflows, both showing similar revenue and margin profiles. Business A has been using AI for eighteen months. Adoption is high. The team is enthusiastic. But workflows are undocumented, measurement is informal, and the leadership team can't produce a structured AI operating record. Business B has been running the AI Operating System from this book for three quarters. It has current Workflow Reality Maps for its five highest-activity workflows, documented baseline and current measurements for all four core metrics, a completed AI Efficiency Scorecard, ROI calculations for each active use case, and a day ninety presentation from each of the three completed optimization cycles. Both businesses are generating similar financial results today. But Business B can prove why its AI operation produces those results, can show it's improving every quarter, and can demonstrate that the improvement is documented, governed, and repeatable. An acquirer evaluating both businesses will see Business A as a business that uses AI and Business B as a business that has built AI into its operating discipline. That difference affects the risk assessment, the confidence in forward projections, and in the end, the price.

Repeatable improvement is more valuable than isolated AI wins (Schiros et al., 2025) . A one-time AI success story may impress people in a meeting. A documented quarterly operating rhythm that shows consistent, measured improvement makes the business look scalable, governable, and easier to trust. That's the standard the AI Operating System is designed to meet. And it's the standard that produces the most durable financial advantage from AI over time.

The Due Diligence AI Operations Package

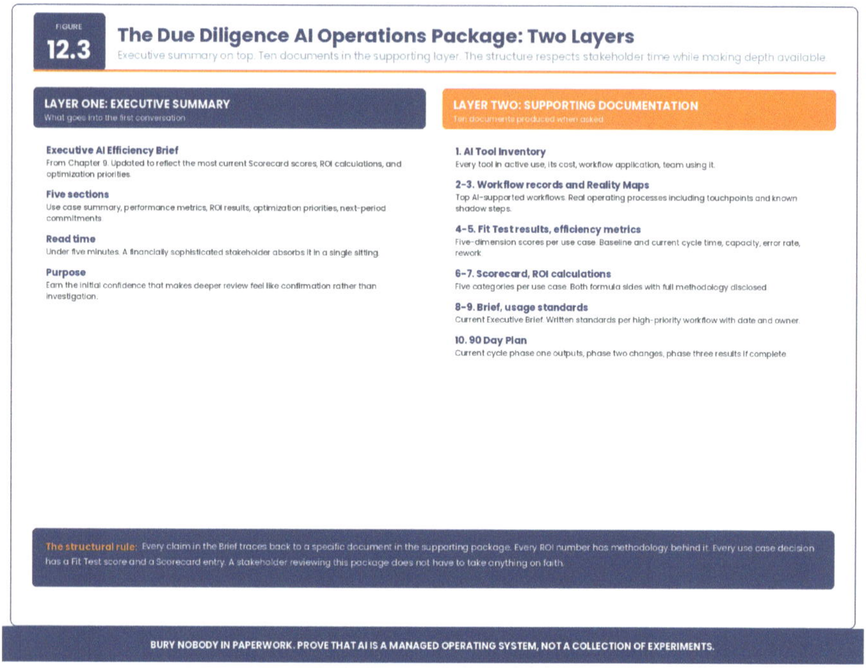

There will be a moment when a financial stakeholder asks to see the documentation behind the AI operations. It might come during an acquisition due diligence process. It might come during a lender's annual review. It might come from a board member who wants to see the evidence behind the AI performance claims in the last quarterly update. It might come from an investor who is evaluating whether to increase their position. When that moment arrives, the business that has assembled a complete, organized AI operations package will handle it with confidence. The business that hasn't will spend the next two weeks scrambling to reconstruct documentation that should have been maintained all along.

The due diligence AI operations package is the organized collection of documentation that proves AI inside the business is visible, measured, governed, standardized, and improving. It's not a single document. It's a structured package with two layers: an executive summary for the first conversation and supporting documentation for deeper review. That two-layer structure is important because it respects the financial stakeholder's time while making depth available when they ask for it.

The executive summary layer is the Executive AI Efficiency Brief from Chapter 9, updated to reflect the most current Scorecard scores, ROI calculations, and optimization priorities. This is the document that goes into the room first. It's one page. It covers the active AI use cases, the efficiency performance metrics, the ROI results, the optimization priorities, and the next-period commitments. Any financially sophisticated stakeholder can read it in under five minutes and understand what AI is doing in the business, what it's returning, and what leadership is doing to improve performance. The Brief is the entry point. It's what earns the initial confidence that makes a deeper review feel like confirmation rather than investigation.

The supporting documentation layer contains ten specific items. The first is the current AI Tool Inventory, showing every AI tool in active use, its cost, its workflow application, and the team or department using it. This document answers the basic question of what AI is being used and what it costs, before anyone has to ask. The second is the documentation of the top AI-supported workflows, showing for each one the workflow name, the AI tool involved, the operational purpose, the team responsible, and the current performance status. The third is the Workflow Reality Maps for the highest-activity workflows, showing the real operating process including AI touchpoints, review steps, handoffs, and any known shadow processes or exception handling routines.

The fourth item is the AI Process Fit Test results for each active use case, showing the scores across all five dimensions, the rationale for each score, the decision assigned, and the date of the most recent re-score. The fifth is the baseline and current efficiency metrics for each workflow where AI is active, showing the pre-AI and current measurements for cycle time, capacity, error rate, and rework cost. The sixth is the AI Efficiency Scorecard, showing the current scores across all five categories for every active use case, the trend direction compared to the previous review period, and the decision recommendation for each use case. The seventh is the AI ROI Formula calculations for each active use case, showing both sides of the formula with full methodology documentation for every input, including the complete cost denominator with the AI Efficiency Tax components itemized.

The eighth item is the most recent Executive AI Efficiency Brief, which serves as the summary of the entire operating record. The ninth is the AI usage standards documentation, showing the written standards for each high-priority workflow, the date they were created or last updated, and the owner responsible for maintaining them. This document demonstrates that AI usage is governed by documented standards rather than individual judgment. The tenth is the current 90 Day AI Process Optimization Plan, showing the three workflows selected for the current cycle, the phase one outputs including baselines and prioritized action list, the phase two changes made and before-and-after measurements, and the phase three results if the current cycle is complete. This document demonstrates that improvement is a structured, recurring discipline rather than an ad hoc response to problems as they appear.

The purpose of assembling this package isn't to bury a buyer, lender, investor, or acquirer in paperwork. It's to prove that AI inside the business is not a collection of tools and experiments. It's a managed, measured, improving operating system with documented evidence at every level. A financial stakeholder who reviews this package doesn't have to take anything on faith. Every claim in the Executive AI Efficiency Brief traces back to a specific document in the supporting package. Every ROI number has a methodology behind it. Every use case decision has a fit test score and a Scorecard entry supporting it. Every improvement claim has a before-and-after measurement record confirming it.

Keep the package organized and current. Assign one person the explicit responsibility for maintaining it, updating it after each 90 Day AI Process Optimization Plan cycle, and confirming that every document in the supporting layer is current before any high-stakes financial conversation. A package that hasn't been updated since the last cycle ended is a package that will create questions rather than answer them. The maintenance discipline is as important as the initial assembly, because a financial stakeholder reviewing the package will notice immediately if the most recent Scorecard update is six months old or if the ROI calculations haven't been refreshed since the first cycle. Currency signals that the operating discipline is running. Stale documentation signals that it ran once and stopped.

Putting It Into Practice

This chapter mapped the connection between AI operating discipline and business valuation. The chapter's work is not complete until the discipline has been assembled into a package the leadership team can put in front of a financial stakeholder. Complete the four-step assembly below on the materials each prior chapter produced.

First, pick the first stakeholder conversation the package will support. An upcoming acquisition discussion, a lender annual review, a board update, or an investor relationship. The first target shapes the emphasis. The data does not change between audiences, but the cover note and the order of the supporting documents do. Pick one. Schedule a date. Work backward from the date to determine what must be ready by when.

Second, pull together the ten supporting items from this chapter into one organized folder: AI Tool Inventory, workflow documentation, Reality Maps, Fit Test results, baseline and current measurements, the Scorecard, the ROI calculations, the most recent Executive Brief, the AI usage standards, and the current 90 Day Optimization Plan. Each item carries a date and an owner. A document without provenance is not part of the package; it is paperwork.

Third, run the brief-first discipline. The Executive Brief goes into the room first. Supporting documents are produced only when a specific question requires them. A financial stakeholder who receives twenty documents upfront and is expected to read them before the meeting is less engaged and less confident than one who receives a clean one-page summary and then receives specific supporting documents in response to specific questions. The Brief opens the door. The package confirms what is behind it.

Fourth, assign one person the explicit responsibility for maintaining the package. After each 90 Day cycle, the Brief, the Scorecard, the ROI calculations, and the standards documentation get refreshed. A package that has not been updated since the last cycle ended will create questions rather than answer them. Currency signals that the operating discipline is running. Stale documentation signals that it ran once and stopped.

AI operating discipline is not just an internal management tool. It is a financial asset. Proof is what this entire book was built to produce, and the proof exists when the package exists. The Conclusion that follows ties the twelve chapters into a single closing statement and lays out the sequenced starting point for any leadership team ready to install the discipline this Volume has installed.

FIGURE 12.4

Putting It Into Practice

Assemble the package every prior chapter has produced. The discipline becomes a financial asset the moment it is documented and put in front of a stakeholder

THE FOUR INPUTS YOU NEED BEFORE ASSEMBLING THE PACKAGE

I1 — The eleven prior chapters' outputs
Tool inventory, workflow docs, Reality Maps, Fit Test results, baselines, Scorecard, ROI worksheets, Brief, standards, 90 Day Plan. The chapter outputs are the package contents.

I2 — The first target conversation
An upcoming acquisition discussion, lender review, board update, or investor conversation. The first target shapes the cover note. Pick one.

I3 — A named package owner
One person responsible for assembly, currency, and the brief-first discipline. Without a named owner, the package becomes everyone's job, which means it becomes no one's job.

I4 — A maintenance cadence
The package gets refreshed after each 90 Day cycle. The cadence is on the leadership calendar before the first version is presented. Currency is what keeps the package as an asset.

1 PICK THE FIRST TARGET CONVERSATION

Pick one stakeholder relationship the package will support first. Acquisition diligence, lender annual review, board update, or investor conversation. Schedule a date. Work backward from the date to determine what must be ready by when. One target. One date. One owner.

Owner: Executive Sponsor with CFO

2 ASSEMBLE THE TEN ITEMS

Pull the ten supporting documents from this chapter into one organized folder. Each item carries a date and an owner. A document without provenance is not part of the package; it is paperwork. The folder is the deliverable, not the meetings about the folder

Owner: AI Efficiency Lead — within fourteen days

3 LEAD WITH THE BRIEF

Run the brief-first discipline. The Executive Brief enters the room first. Supporting documents are produced only when a specific question requires them. The Brief opens the door. The package confirms what is behind it. Twenty documents upfront produce confusion, not confidence.

Owner: Executive Sponsor — in every external conversation

4 ASSIGN MAINTENANCE AND CALENDAR UPDATES

Assign one person the explicit responsibility for maintaining the package. After each 90 Day cycle, the Brief, Scorecard, ROI, and standards documentation get refreshed. Currency signals that the discipline is running. Stale documentation signals that it ran once and stopped.

Owner: Named Package Owner — calendar entry per 90 Day cycle

PROOF IS WHAT THIS BOOK WAS BUILT TO PRODUCE. THE PROOF EXISTS WHEN THE PACKAGE EXISTS.

Conclusion

What This Book Was Really About

This book was never a book about AI tools. It wasn't written to help you discover new platforms, build excitement around adoption, or expand the number of employees logging into AI software each week. Every chapter, every framework, every measurement method in this volume was built around a single operating truth that most businesses haven't fully confronted yet: AI activity is not AI efficiency (Conant et al., 2019). Those two things look identical from a distance and produce completely different financial results up close.

The AI Efficiency Gap isn't a technology problem. It's a measurement problem. And measurement problems don't fix themselves (Wainer, 1993).

From the moment the AI Efficiency Gap was introduced in Chapter 1, the book was asking one question in different ways. Not "are your employees using AI?" but "has AI actually moved anything in this business that you can defend with a number?" Cycle time. Capacity. Error rates. Rework. Labor savings. Margin. Those are the categories where AI performance either shows up or doesn't. And until a business measures those categories specifically, it's managing a story rather than a result. A story might satisfy a leadership meeting. It won't satisfy a CFO, a board member, a lender, or an acquirer who is asking for evidence before making a financial decision.

Phantom Productivity showed you what it looks like when AI creates the appearance of efficiency without the substance of it. The AI Efficiency Tax showed you what that appearance costs in hidden labor, review burden, rework cycles, and management overhead that nobody is tracking as an AI-related expense. The Workflow Reality Map showed you the gap between the workflow leadership thinks is running and the workflow employees are actually running (Retelny et al., 2017) every day. The AI Process Fit Test gave you a structured way to determine whether AI belongs where it's been placed, or whether the business is investing in a tool that's accelerating a broken process rather than improving a functional one.

The measurement work in Chapter 6 gave you the four indicators that actually prove performance: cycle time, capacity, error rate, and rework. The AI Efficiency Scorecard organized those indicators into a single operating view that any leadership team can review, compare, and act on in one sitting. The AI ROI Formula converted those measurements into a financial return figure that can hold up in a boardroom without falling apart when someone asks a follow-up question. The Executive AI Efficiency Brief packaged all of it into the format financial stakeholders actually need. And the 90 Day AI Process Optimization Plan gave you the sequenced execution structure to turn all of that evidence into documented, recurring operational improvement.

Chapter 10 on standardization and Chapter 12 on business valuation weren't additions to the operating system. They were the completion of it. Standardization is what makes improvement repeatable (Bakhtiari, 2026) . Valuation is what happens when repeatable improvement is documented well enough that financial stakeholders can see it, trust it, and price it into their decisions. The entire book was always heading toward that outcome. An AI operation that can be seen in the workflows, defended in the boardroom, and reflected in the P&L.

That's what this book was really about.

The Action Is the Outcome

There's a difference between a leader who read this book and a leader who used it. That difference is the only one that matters now.

Reading produces understanding. Using produces results. (Locke et al., 2009) And the results this book was designed to produce don't come from understanding the AI Efficiency Gap concept. They come from mapping the three workflows where the gap is widest, measuring what's actually running inside them, and making a documented decision about what to change. They don't come from knowing what Phantom Productivity looks like. They come from running the downstream measurement that proves whether AI is creating real efficiency or only visible motion in your specific workflows. They don't come from understanding the AI Efficiency Tax in principle. They come from calculating the monthly dollar figure for your highest-activity use case and putting that number in front of your CFO.

The businesses that will build defensible AI operating systems over the next several years won't be the ones waiting for better tools, more data, or the perfect conditions to start measuring. They'll be the ones that started the first 90 Day AI Process Optimization Plan with the workflows they already had, the measurements they could capture this week, and the evidence they were willing to build honestly even when the early numbers were uncomfortable.

Uncomfortable numbers are more valuable than comfortable stories (Pandey et al., 2021) . A negative AI ROI on a use case that's been running for a year is not a failure. It's information. It tells the business exactly where to focus the next optimization cycle, which investment to redirect, and which conversation to have with the team before the problem compounds further. A business that can find that number, report it honestly, and act on it quickly is a business that's managing AI with operating discipline. That discipline is the competitive advantage. Not the tools. Not the adoption rate. The discipline.

The frameworks in this book don't expire. The AI Efficiency Scorecard doesn't become less useful as tools evolve. The Workflow Reality Map doesn't become less accurate as workflows change. The AI ROI Formula doesn't become less defensible as financial expectations rise. If anything, each of these tools becomes more valuable as the business runs more cycles through the 90 Day AI Process Optimization Plan, because the measurement record gets more detailed, the baselines get sharper, and the improvement story gets stronger with every quarter of documented results behind it.

The decision sitting in front of every leader who has reached this page is straightforward. AI either remains a story inside the business, or it becomes a measurable operating advantage. The frameworks are in hand. The measurement system is described. The execution sequence is laid out. What converts all of that into a result is the decision to begin, and the discipline to continue past the first ninety days into the second cycle, the third, and every quarter after that where the operating discipline compounds and the financial case for AI gets cleaner, stronger, and harder to challenge.

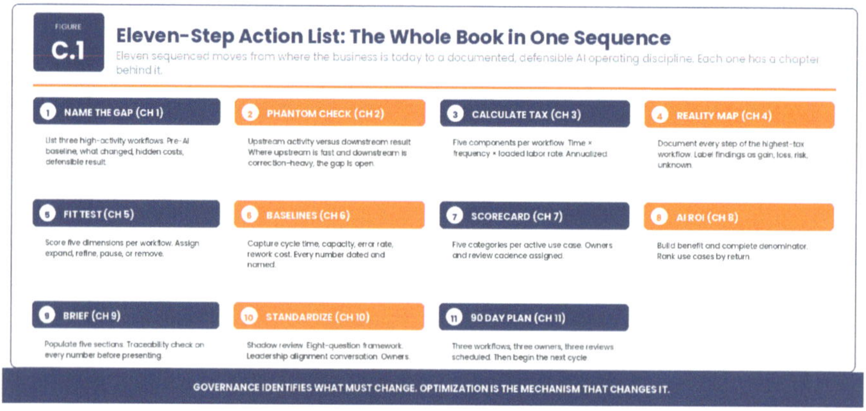

Your Step-by-Step Starting Point. The Whole Book in One Action List

Eleven sequenced moves from where the business is today to a documented, defensible AI operating discipline. Each one has a chapter behind it.

1. Name the gap. List the three AI-supported workflows where activity is highest. For each, write down the pre-AI baseline, what measurably changed, what hidden costs appeared, and what financial result you can defend today. (Chapter 1)

2. Run the Phantom Productivity check. Map upstream activity against downstream result for each workflow. Where the upstream looks fast and the downstream looks slow and correction-heavy, you have found it. (Chapter 2)

3. Calculate the AI Efficiency Tax. For each workflow, estimate the five components: review burden, rework cycles, output inconsistency cost, management overhead, and shadow process cost. Time times frequency times loaded labor rate, totaled and annualized. (Chapter 3)

4. Build the Workflow Reality Map. For the highest-tax workflow, document every step, handoff, AI touchpoint, workaround, and informal approval. Label findings as gain, loss, risk, or unknown. (Chapter 4)

5. Run the AI Process Fit Test. Score five dimensions per workflow. Assign one of four decisions: expand, refine, pause, or remove. (Chapter 5)

6. Establish baselines for the four core performance indicators. Capture cycle time, capacity, error rate, and rework cost across each selected workflow. Every number dated and named. (Chapter 6)

7. Build the AI Efficiency Scorecard. Score each active use case across five categories. Assign owners and a recurring review cadence. (Chapter 7)

8. Calculate AI ROI. Build the benefit side from labor savings, capacity gains, error and rework cost eliminated, and Tax reduction. Build the complete cost denominator. Rank the use cases by return. (Chapter 8)

9. Draft the Executive AI Efficiency Brief. Populate the five sections. Run the traceability check on every number before presenting. (Chapter 9)

10. Standardize AI usage. Conduct the shadow process review, draft usage standards using the eight-question framework, hold the leadership alignment conversation. Assign owners. (Chapter 10)

11. Launch the first 90 Day AI Process Optimization Plan. Three workflows, three named owners, three scheduled reviews on the calendar before day one. Then begin the next cycle. (Chapter 11)

Governance identifies what must change. Optimization is the mechanism that changes it. That is what this book installs. The starting point is the first workflow on the list, and the first measurement captured this week.

Framework Alignment

Where the operating measurement, financial integration, and valuation work in this Volume connects to external frameworks

Volume III completed the formal framework alignment work for the Library, mapping every governance instrument to ISO/IEC 42001, the NIST AI Risk Management Framework, the EU AI Act, NYC Local Law 144, SR 11-7, and the sector-specific rules. Volume IV does not repeat that alignment. It inherits it. The optimization instruments in this Volume support continued alignment with the same frameworks Volume III mapped, because the measurement, scoring, and ROI work this Volume produces becomes the operating evidence that the framework clauses on monitoring, measurement, and continual improvement expect to find in place.

This section identifies the specific frameworks and authoritative sources that the Volume IV operating instruments most directly support, organized by the type of alignment each one represents.

ISO/IEC 42001 Monitoring, Measurement, and Continual Improvement

Clause 9.1 (monitoring, measurement, analysis, and evaluation) is the clause that the AI Efficiency Scorecard™, the cycle time and capacity measurements in Chapter 6, and the AI ROI Formula directly support. The Scorecard is the documented evidence that the business is monitoring the performance of its AI use cases on a consistent cadence with named owners and recorded decisions.

Clause 10.2 (continual improvement) is the clause that the 90 Day AI Process Optimization Plan supports. The recurring cycle is the operating evidence that the business is improving its AI performance through deliberate, documented action rather than incidental change.

NIST AI Risk Management Framework: MEASURE and MANAGE

The MEASURE function (specifically MEASURE 2 on trustworthiness characteristics) and the MANAGE function (specifically MANAGE 1 on risk prioritization and MANAGE 4 on risk treatment and recovery) are the functions that the Volume IV optimization work most directly supports. The Scorecard provides documented MEASURE evidence. The 90 Day Plan and the AI Process Fit Test provide documented MANAGE evidence. The integration of governance findings from Volume III with the optimization actions in Volume IV is the bridge that the NIST Framework expects between identifying risk and treating it.

Financial Reporting and Audit Standards

FASB ASU 2025-06, Targeted Improvements to the Accounting for Internal-Use Software (Subtopic 350-40), issued by the Financial Accounting Standards Board in September 2025, governs when entities that internally develop software begin capitalizing those costs. ASU 2025-06 may apply to organizations that internally develop AI software; it does not generally change the accounting treatment for organizations that consume AI through third-party SaaS subscriptions, which is the more common pattern for mid-market firms. The AI ROI Formula's cost denominator described in this Volume is consistent with the general cost categories financial professionals use to evaluate technology investments, but the AI ROI Formula is an operating instrument for management use, not a financial reporting standard. Businesses with material AI investments should consult their accounting professionals on the specific application of ASU 2025-06 and related FASB guidance to their financial statements.

AICPA AI Implementation Checklist (July 2025) and SAS 146 (Quality Management for an Engagement, effective December 15, 2025) are the audit standards most relevant to the operating evidence Volume IV produces. The Scorecard, the workflow reality maps, and the documented baseline measurements are the kind of operating evidence that external auditors increasingly expect to see in place when AI affects materially-reported numbers.

Operating Discipline Frameworks and Methodologies

Lean operating principles on baseline measurement, cycle time analysis, and rework cost identification are the operating discipline foundations the Volume IV measurement work draws from. The methodology is adapted for AI workflows, where the unit of analysis is the AI-assisted process rather than the manufacturing line, but the underlying discipline of measuring before changing, capturing baselines honestly, and treating activity metrics as separate from performance metrics is consistent with lean practice.

Standard SMB management accounting practice on loaded labor rate calculation, contribution margin analysis, and operating expense classification underpins the financial calculations throughout the Volume. The methodology in Chapter 3 on calculating the loaded labor rate in an SMB context is adapted from standard management accounting practice for the specific case of measuring AI-related labor cost.

Business Valuation and Diligence Frameworks

Standard small and mid-sized business valuation practice on quality-of-earnings analysis, operating add-back identification, and diligence package preparation provides the framework for the Volume IV work on AI's effect on business valuation. The Due Diligence AI Operations Package in Chapter 12 is built from the diligence categories that quality-of-earnings practitioners and acquisition diligence teams expect to see populated. Businesses preparing for a lending, investment, or acquisition event should consult their financial and legal advisors on the specific application of these frameworks to their situation.

Continuing Alignment with Volume III

The full regulatory and standards alignment for the Library is documented in Volume III's Framework Alignment section. That alignment continues to apply to the Volume IV work. The measurement, scoring, and ROI evidence this Volume produces strengthens the framework alignment Volume III established, because operating evidence that is current, dated, documented, and reviewable on a consistent cadence is what every framework on the list expects to find when an auditor or examiner looks at the AI program in operation.

References

This References section consolidates the regulations, standards, authoritative pronouncements, scholarly works, and case law cited throughout this Volume. Entries are presented in American Psychological Association (APA) 7th edition format and listed alphabetically by first author or issuing body. Where multiple works by the same author or issuer appear in the same year, alphabetic suffixes (a, b, c) distinguish them. The complete regulatory and standards reference set for the Library is in Volume III.

Acemoglu, D., & Restrepo, P. (2018). Artificial intelligence, automation, and work. In *The economics of artificial intelligence: An agenda* (pp. 197-236). University of Chicago Press.

Agha-Mir-Salim, L., Alberto, I. R., Alberto, N. R., Celi, L. A., Alfonso, P. G., Hicklen, R., ... & Sauer, C. M. (2025). Technological solutions to improve inpatient handover in the era of artificial intelligence: scoping review. *Journal of Medical Internet Research*, 27, e70358.

Agyei, K., Nemure, T., Gwangwava, S., Chisora, H. H., Samushonga, C., Monjoma, M. B., & Mupa, M. N. (2026). Governance-driven security automation for municipal and SMB cloud modernization: A NIST CSF 2.0–Aligned Remediation Operating Model. *World Journal of Advanced Research and Reviews*, 30(1), 2064-2073.

Akpan, M. (2025). Will AI Lead to Abundance? Exploring Cost Reductions from Streaming to Tokenized Technologies. *Finance Research Open*, 100043.

Anderson, E. W., & Fornell, C. (2000). Foundations of service quality as a leading indicator. *Handbook of services marketing and management*, 255.

Arashpour, M., & Arashpour, M. (2015). Analysis of workflow variability and its impacts on productivity and performance in construction of multistory buildings. *Journal of Management in Engineering*, 31(6), 04015006.

Arora, J. S., & Baenziger, G. (1986). Uses of artificial intelligence in design optimization. *Computer Methods in Applied Mechanics and Engineering*, 54(3), 303-323.

Babar, Z., Paul, R., Rahman, M. A., & et al. (2025). A systematic review of human-AI collaboration in IT support services: enhancing user experience and workflow automation. *Journal of Human-AI Interaction & Automation*, 11(2), 142-158.

Bakhtiari, M. (2026). Standardization Versus Adaptability: Where Is the Sweet Spot?. *Practical Radiation Oncology*.

Bharadwaj, P., Nicola, L., Breau-Brunel, M., Sensini, F., Tanova-Yotova, N., Atanasov, P., ... & Blankenburg, M. (2024). Unlocking the value: quantifying the return on investment of hospital artificial intelligence. *Journal of the American College of Radiology*, 21(10), 1677-1685.

Bishop, J. M. (2021). Artificial intelligence is stupid and causal reasoning will not fix it. *Frontiers in Psychology*, 11, 513474.

Blank, G. (2006). *Critics, ratings, and society: The sociology of reviews*. Bloomsbury Publishing USA.

Blasch, E., Sung, J., & Nguyen, T. (2021). Multisource AI scorecard table for system evaluation. *arXiv preprint arXiv:2102.03985*.

Bly, R. (2009). *The complete guide to money making writing jobs*.

Boediman, A., & Ferdiana, R. (2025). *AI FALL: Why may AI be smarter than us but still lack common sense?*. Jawara Press.

Bosse, G., Breuer, J. P., & Spies, C. (2006). The resistance to changing guidelines–what are the challenges and how to meet them. *Elsevier Systems and Controls*.

Braithwaite, J., & Makkai, T. (1994). Trust and compliance. *Policing and Society: An International Journal*, 4(1), 1-12.

Brownstone, D. (1996). Using percentage accuracy to measure neural network predictions in stock market movements. *Neurocomputing*, 10(3), 237-250.

Brynjolfsson, E., Rock, D., & Syverson, C. (2018). Artificial intelligence and the modern productivity paradox: A clash of expectations and statistics. In *The economics of artificial intelligence: An agenda* (pp. 23-57). University of Chicago Press.

Buckley, R. P., Zetzsche, D. A., Arner, D. W., & Tang, B. W. (2021). Regulating artificial intelligence in finance: Putting the human in the loop. *Sydney Law Review, The*, 43(1), 43-81.

Bughin, J., Hazan, E., Sree Ramaswamy, P., DC, W., & Chu, M. (2017). *Artificial intelligence: The next digital frontier*. McKinsey Global Institute.

Burgess, A. (2018). *The executive guide to artificial intelligence. How to identify and implement applications for AI in your organization*. AJBurgess Ltd.

CHANG, W. W. V. (2020). Labor Displacement in Artificial Intelligence Era: A Systematic Literature Review. *Taiwan Journal of East Asian Studies*, 17(2).

Checco, A., Bracciale, L., Loreti, P., Pinfield, S., & Bianchi, G. (2021). AI-assisted peer review. *Humanities and Social Sciences Communications*, 8(1), 25.

Chui, M., & Francisco, S. (2017). Artificial intelligence the next digital frontier. *McKinsey and Company Global Institute*, 47(3.6), 6-8.

Cokins, G. (2004). *Performance management: Finding the missing pieces (to close the intelligence gap)*. John Wiley & Sons.

Conant, E. F., Toledano, A. Y., Periaswamy, S., Fotin, S. V., Go, J., Boatsman, J. E., & Hoffmeister, J. W. (2019). Improving accuracy and efficiency with concurrent use of artificial intelligence for digital breast tomosynthesis. *Radiology: Artificial Intelligence*, 1(4), e180096.

Dafoe, A. (2018). AI governance: a research agenda. *Governance of AI Program, Future of Humanity Institute, University of Oxford*, 1442-1443.

Demerjian, P. R. (2017). Uncertainty and debt covenants. *Review of Accounting Studies*, 22(3), 1156-1197.

Deng, L. (2018). Artificial intelligence in the rising wave of deep learning: The historical path and future outlook [perspectives]. *IEEE Signal Processing Magazine*, 35(1), 180-177.

Devarajan, H., Pottier, L., Velusamy, K., Zheng, H., Yildirim, I., Kogiou, O., ... & Mohror, K. (2024, November). Dftracer: An analysis-friendly data flow tracer for AI-driven workflows. In *SC24: International Conference for High Performance Computing, Networking, Storage and Analysis* (pp. 1-24). IEEE.

Dignam, A. J. (2019). Artificial intelligence: The very human dangers of dysfunctional design and autocratic corporate governance. *Queen Mary School of Law Legal Studies Research Paper*, (314).

Drew, S. A., Kelley, P. C., & Kendrick, T. (2006). CLASS: Five elements of corporate governance to manage strategic risk. *Business Horizons*, 49(2), 127-138.

Ellis, C. A., Keddara, K., & Rozenberg, G. (1995). Dynamic change within workflow systems. *ACM SIGOIS Bulletin*, 16(1), 10-14.

Fischer, L., Ehrlinger, L., Geist, V., Ramler, R., & et al. (2020). AI system engineering—key challenges and lessons learned. *Machine Learning and Knowledge Extraction*, 3(1), 4-18.

Friebe, M. (2026). AI in radiology and interventions: a structured narrative review of workflow automation, accuracy, and efficiency gains of today and what's coming. *International Journal of Computer Assisted Radiology and Surgery*, 21(1), 1-10.

Gates, A., Ali, A., Conard, S., & Dunn, P. (2026). Exploring an AI-First Healthcare System. *Bioengineering*, 13(1), 112.

George, A. S. (2024). Artificial intelligence and the future of work: job shifting not job loss. *Partners Universal Innovative Research Publication*, 2(2), 17-37.

Giudici, P. (2018). Fintech risk management: A research challenge for artificial intelligence in finance. *Frontiers in Artificial Intelligence*, 1, 1-12.

Goderis, A., Sattler, U., Lord, P., & Goble, C. (2005). Seven bottlenecks to workflow reuse and repurposing. In *The Semantic Web: Research and Applications* (pp. 400-414). Springer Berlin Heidelberg.

Graham, I., Goodall, P., Peng, Y., Palmer, C., West, A., Conway, P., ... & Dettmer, F. U. (2015). Performance measurement and KPIs for remanufacturing. *Journal of Remanufacturing*, 5(1), 10.

Hanes, S. (2012). *Canadian tourism SMEs: understanding the motivations, valuations of success and experiences of business owners in Southern Ontario.*

Haque, E. A. (2025). The Role of Calibration Engineering In Strengthening Reliability Of US Advanced Manufacturing Systems Through Artificial Intelligence. *Review of Applied Science and Technology*, 4(02), 820-851.

Hasan, R. (2025). A Systematic Review Of Human-AI Collaboration In IT Support Services: Enhancing User Experience And Workflow Automation. *American Journal of Information Systems Research.*

Henry, G. T. (1990). *Practical sampling* (Vol. 21). Sage.

Hoitash, R., Hoitash, U., & Kurt, A. C. (2016). Do accountants make better chief financial officers?. *Journal of Accounting and Economics*, 61(2-3), 414-432.

Iansiti, M., & Lakhani, K. R. (2020). *Competing in the age of AI: Strategy and leadership when algorithms and networks run the world*. Harvard Business Press.

Igwe-Nmaju, C., & Anadozie, C. (2022). Commanding digital trust in high-stakes sectors: communication strategies for sustaining stakeholder confidence amid technological risk. *World Journal of Advanced Research and Reviews*, 15(3), 609-630.

Janzen, K. A. (2025). *Inside AI Transition: Organizational Change in the Making: A Study of Early-Stage AI Implementation in Transitioning SMEs.*

Jennings, L. (2024). *The AI Edge: Unlocking Profits with Artificial Intelligence.* eBookIt.com.

Kammer, P. J., Bolcer, G. A., Taylor, R. N., Hitomi, A. S., & Bergman, M. (2000). Techniques for supporting dynamic and adaptive workflow. *Computer Supported Cooperative Work (CSCW)*, 9(3), 269-292.

Kamraju, M. (2025). The Artificial Intelligence Bubble: An Interdisciplinary Analysis of Economic Overvaluation, Public Perception, and Technological Reality. *Journal of Third World Economics (JTWE)*, 3(2), 54-62.

Kerzner, H. (2025). *Project management: a systems approach to planning, scheduling, and controlling*. John Wiley & Sons.

Khalil, M. A., Hadid, M., Padmanabhan, R., Elomri, A., & Al-Hammadi, Y. (2025). An integrated artificial intelligence and optimization model for operational efficiency and risk reduction in letter of credit examination process. *Future Generation Computer Systems*, 164.

Kumar, M. (2023). Bridging the gap: Integrating AI assisted code suggestions in your UI workflow. *International Journal of Engineering Technology Research & Management (IJETRM)*, 9(03), 71-80.

Lee, J. T., Shen, T. K. B., Liu, V. T. N., Chen, T. H. H., Lee, C. C., Lu, C., ... & Atun, R. (2025). Costing Methods for Artificial Intelligence: Systematic Review and Recommended Cost Inventory for Health Technology Assessment. *medRxiv*, 2025-10.

Lipscomb, T. (2011). *Re-made in the USA: how we can restore jobs, retool manufacturing, and compete with the world*. John Wiley & Sons.

Locke, L. F., Silverman, S. J., & Spirduso, W. W. (2009). *Reading and understanding research*. Sage.

Luecke, R. (2003). *Managing change and transition* (Vol. 3). Harvard Business Press.

Maister, D. H. (1982). Balancing the professional service firm. *Sloan Management Review*, 24(1), 15-29.

Major, A., & Ngalamou, L. (2023, October). Improving Accuracy and Efficiency in Document Handling for Business Processes. In *Proceedings of the Future Technologies Conference* (pp. 543-562). Springer Nature Switzerland.

Mallmann, G. L., Maçada, A. C. G., & Oliveira, M. (2018). The influence of shadow IT usage on knowledge sharing: An exploratory study with IT users. *Business Information Review*, 35(1), 17-28.

Meyer, M. W. (2002). Finding performance: The new discipline in management. *Business Performance Measurement: Theory and Practice*, 51-62.

Milchanowski, K. L. (2026). *Return on Intelligence: A Strategic Enterprise Playbook for Scalable AI Agents*. Taylor & Francis.

Mittman, B. S. (2004). Creating the evidence base for quality improvement collaboratives. *Annals of Internal Medicine*, 140(11), 897-901.

Mohseni, S., Zarei, N., & Ragan, E. D. (2021). A multidisciplinary survey and framework for design and evaluation of explainable AI systems. *ACM Transactions on Interactive Intelligent Systems (TiiS)*, 11(3-4), 1-45.

Moro-Visconti, R. (2024a). *Artificial intelligence valuation*. Springer Nature.

Moro-Visconti, R. (2024b). Artificial intelligence-driven digital scalability and growth options. In *Artificial intelligence valuation: The impact on automation, biotech, chatbots, fintech, B2B2C, and other industries* (pp. 131-204). Springer Nature Switzerland.

Mueller, D. C., & Reardon, E. A. (1993). Rates of return on corporate investment. *Southern Economic Journal*, 60(2), 430-453.

Mueller, J. P., & Massaron, L. (2021). *Artificial intelligence for dummies*. John Wiley & Sons.

Møller, N. L. H., Bjørn, P., Villumsen, J. C., & Bødker, S. (2017). Data tracking in search of workflows. In *Proceedings of the 2017 CHI Conference on Human Factors in Computing Systems* (pp. 2017-2028). ACM.

Narayanan, A., & Kapoor, S. (2025). *AI snake oil: What artificial intelligence can do, what it can't, and how to tell the difference*. Princeton University Press.

Ng, I. C., Wirtz, J., & Sheang Lee, K. (1999). The strategic role of unused service capacity. *International Journal of Service Industry Management*, 10(2), 211-244.

Oke, S. A. (2008). A literature review on artificial intelligence. *International Journal of Information and Management Sciences*, 19(4), 535-570.

Ooi, V., & Goh, G. (2018). *Taxation of automation and artificial intelligence as a tool of labour policy*. National University of Singapore Research Series.

Pandey, S., Gupta, S., & Chhajed, S. (2021). ROI of AI: Effectiveness and measurement. *International Journal of Engineering Research & Technology (IJERT)*, 10(6), 412-428.

Park, H., Ahn, D., Hosanagar, K., & Lee, J. (2021, May). Human-AI interaction in human resource management: Understanding why employees resist algorithmic evaluation at workplaces and how to mitigate burdens. In *Proceedings of the 2021 CHI Conference on Human Factors in Computing Systems* (pp. 1-15).

Patel, S., Shaw, P., & Dale, B. G. (2001). Set-up time reduction and mistake proofing methods–A study of application in a small company. *Business Process Management Journal*, 7(1), 65-75.

Pierre, K., Haneberg, A. G., Kwak, S., Peters, K. R., Hochhegger, B., Sananmuang, T., ... & Forghani, R. (2023, April). Applications of artificial intelligence in the radiology roundtrip: process streamlining, workflow optimization, and beyond. In *Seminars in Roentgenology* (Vol. 58, No. 2, pp. 158-169). WB Saunders.

Piscitelli, A., David, C., De Rosa, M., Mohammed, A., Nanni, F., Pake, J., ... & Zheng, C. (2025). AI-Assisted Authoring for Transparent, Data-Driven Documents. *arXiv preprint arXiv:2601.06027.*

Prati, A., Mikic, I., Trivedi, M. M., & Cucchiara, R. (2003). Detecting moving shadows: algorithms and evaluation. *IEEE Transactions on Pattern Analysis and Machine Intelligence*, 25(7), 918-923.

Prescott, L. D. (2024). *AI Agents: The Invisible Workforce of the Future*. eBo okIt.com.

Quaquebeke, N. V., & Gerpott, F. H. (2023). The now, new, and next of digital leadership: How Artificial Intelligence (AI) will take over and change leadership as we know it. *Journal of Leadership & Organizational Studies*, 30(3), 265-275.

Ranschaert, E., Topff, L., & Pianykh, O. (2021). Optimization of radiology workflow with artificial intelligence. *Radiologic Clinics*, 59(6), 955-966.

Retelny, D., Bernstein, M. S., & Valentine, M. A. (2017). No workflow can ever be enough: How crowdsourcing workflows constrain complex work. *Proceedings of the 2017 ACM Conference on Computer Supported Cooperative Work and Social Computing*, 109-122.

Robert, L. P., Pierce, C., Marquis, L., Kim, S., & Alahmad, R. (2020). Designing fair AI for managing employees in organizations: a review, critique, and design agenda. *Human–Computer Interaction*, 35(5-6), 545-575.

Rosenlund, P. (2025). *Strategic communication in the age of AI: from the perspective of senior communication professionals.* Corporate Communication Forum.

Safapour, E., & Kermanshachi, S. (2019). Identifying early indicators of manageable rework causes and selecting mitigating best practices for construction. *Journal of Management in Engineering*, 35(2), 04018060.

Schiros, C., Kaur, S., Arora, R., & Jagannathan, U. (2025). *The AI Optimization Playbook: Drive business success with proven AI strategies, best practices, and responsible innovation*. Packt Publishing.

Schrage, M., Kiron, D., Candelon, F., Khodabandeh, S., & Chu, M. (2023). AI is helping companies redefine, not just improve, performance. *MIT Sloan Management Review*, 64(3), 1-7.

Semionovas, T. (2026). *Governing shadow artificial intelligence in organizations* (Doctoral dissertation, Kaunas University of Technology).

Silic, M., Silic, D., & Kind-Trüller, K. (2025). From shadow IT to shadow AI–threats, risks and opportunities for organizations. *Strategic Change*, 34(1).

Singireddy, J. (2025). *Smart Finance: Harnessing Artificial Intelligence to Transform Tax, Accounting, Payroll, and Credit Management for the Digital Age*. Deep Science Publishing.

Sirota, D., & Klein, D. (2013). *The enthusiastic employee: How companies profit by giving workers what they want*. FT Press.

Solomon, O., Emmanuel, F., Simon, J., & Mendez, D. (2024). Applying advanced analytics, interactive intelligence dashboards, and AI powered predictive models to accelerate entrepreneurial growth, sharpen public sector decision making, and lift organizational profitability. *International Journal of Science, Architecture, Technology, and Environment*, 1(9), 111-129.

Soofastaei, A. (2025). Intelligent scheduling: how AI and advanced analytics are revolutionizing time optimization. In *Mastering Time-Innovative Solutions to Complex Scheduling Problems*. IntechOpen.

Strahan, P. E. (1999). Borrower risk and the price and nonprice terms of bank loans. *Federal Reserve Bank of New York Economic Policy Review*, 5(4).

Strong, D. M., & Miller, S. M. (1995). Exceptions and exception handling in computerized information processes. *ACM Transactions on Information Systems (TOIS)*, 13(2), 206-233.

Turner, V. B. (1919). Wages and hours of labor. *Monthly Labor Review*, 8(3), 102-118.

Umoga, U. J., Sodiya, E. O., Ugwuanyi, E. D., Jacks, B. S., Lottu, O. A., Daraojimba, O. D., & Obaigbena, A. (2024). Exploring the potential of AI-driven optimization in enhancing network performance and efficiency. *Magna Scientia Advanced Research and Reviews*, 10(1), 368-378.

Vasif Aliyev, V. (2026). *From Code to Business Value: The Strategic Role of Full-Stack Development in Digital Startup Success*. ResearchGate.

Vale, S. D. (2025). *The Human Algorithm: Resilient Leadership in AI's Shadow*. eBookIt.com.

Wainer, H. (1993). Measurement problems. *Journal of Educational Measurement*, 30(1), 1-21.

Wamba, S. L. (2020). Influence of artificial intelligence (AI) on firm performance: the business value of AI-based transformation projects. *Business Process Management Journal*, 26(7).

Wang, P. (2019). On defining artificial intelligence. *Journal of Artificial General Intelligence*, 10(2), 1-37.

Webber, S. S., Detjen, J., MacLean, T. L., & Thomas, D. (2019). Team challenges: Is artificial intelligence the solution?. *Business Horizons*, 62(6), 741-750.

Wehenkel, L., Van Cutsem, T., & Ribbens-Pavella, M. (1989). An artificial intelligence framework for online transient stability assessment of power systems. *IEEE Transactions on Power Systems*, 4(2), 789-800.

Weiner, J. (2020). *Why AI/data science projects fail*. Morgan & Claypool Publishers.

Xu, X. G. (2014). An exponential growth of computational phantom research in radiation protection, imaging, and radiotherapy: a review of the fifty-year history. *Physics in Medicine & Biology*, 59(18), R233-R302.

Zhang, L. (2025). AI Waiting Taxes-The Hidden Cost of Idle Time in the Age of Artificial Intelligence. *SSRN Electronic Journal*, papers.ssrn.com.

Zheng, X., & Fan, H. (2024). Exploration of the effectiveness and experience of AI-assisted academic reading. *Proceedings of the Association for Information Science and Technology*, 61(1), 447-456.

Continuing the Work

The AI Efficiency & Process Optimization™ is the fourth Volume in The Operating Discipline for AI Library™, a nine-volume canon on running AI as a permanent business function. Each Volume addresses one of the disciplines a leadership team has to install, in roughly the order a serious organization needs them. The canon is organized into two pillars. The first, AI Business Services™, takes a leadership team from its first audit through full operating maturity and closes with this Volume. The second, AI Risk Governance & Security™, covers the protection and exposure work, including the three security disciplines that AI has structurally changed.

The full canon is below. Each title is also a service line at SRJ Consulting & Services, paired with the working templates and frameworks that move the discipline from a printed page into an operating rhythm.

Pillar I: AI Business Services™

Volume I. The AI Business Enablement Audit™. The diagnostic foundation of the Library. A structured, plain-English evaluation of how AI is operating across a business, what it is costing fully loaded, and whether it is producing measurable outcomes.

Volume II. The AI Readiness & Performance Assessment™. A forward-looking evaluation of whether the organization is prepared to scale AI responsibly, covering workflow maturity, data reliability, internal controls, and the measurement infrastructure AI depends on.

Volume III. The AI Risk & Governance Review™. AI introduces risk at the data, decision, and vendor layers. This Volume establishes governance structures proportionate to the organization's size and risk profile, without slowing the operations that depend on AI.

Volume IV. The AI Efficiency & Process Optimization™. The Volume you are holding. Once AI is understood and governed, this Volume turns to operational effectiveness. Repeatable processes, reduced manual effort, measurable performance, and labor savings that are genuinely captured. The closing Volume of Pillar I.

Pillar II: AI Risk Governance & Security™

Volume V. The AI IT Security Audit™. A CISO-grade audit framework for proving AI exposure is known, controlled, and governed. An exposure map, a remediation roadmap, a regulatory crosswalk, and a board briefing across the six security domains.

Volume VI. The AI IT Security Implementation & Strategy™. The operating manual for proving AI risk is governed. Ratified policies, an integrated risk register, a regulatory crosswalk, and board reporting that survives the question. A discipline to operate, not a project to complete.

Volume VII. Secure by Design in the Age of AI™. Engineering velocity is up an order of magnitude. Security review capacity is not. This Volume defines the Dual-Impedance Problem and the five-framework operating model that closes it for organizations shipping AI-enabled products.

Volume VIII. Application Security in the Age of AI™. Every application security tool ever built was built on the premise that an application produces the same output for the same input. AI invalidated that. This Volume defines the Runtime Determinism Gap and the program model for application security in a world where applications no longer behave deterministically.

Volume IX. Cloud and Infrastructure Security in the Age of AI™. Cloud security was built on the premise that every meaningful action could be traced to an accountable human. AI broke that. This Volume defines the Sovereignty Problem and the cloud security operating model for an era where most actors inside the infrastructure are not human.

Together, these nine Volumes form a complete operating discipline. They are designed to be read in order by a leadership team that is serious about installing AI as a managed function inside the business, and to be referenced by name as the work moves from audit to assessment to governance to optimization to security. Each Volume stands on its own. Read together, they form the Library.

If You Want to Continue This Work

A leadership team that has read this Volume can run the methodology on its own. The frameworks are written to be self-applied. The 90 Day AI Process Optimization Plan was designed for a lean leadership team without outside consulting support, and most of the work in this book can be completed by an internal team with no engagement beyond reading.

For teams that want support running the first cycle, building out the working instruments into real engagement artifacts, or applying the methodology across multiple business units in parallel, advisory work is available through SRJ Consulting and Services. The engagement formats that fit the methodology most naturally are a structured first-cycle facilitation, an AI Efficiency Tax assessment focused on a small set of high-activity workflows, and ongoing quarterly cadence support for leadership teams installing the recurring discipline as a permanent operating function.

The work itself is the point. Whether it is done internally or with help, what matters is that AI inside the business stops being a story and starts being a measured operating discipline. The frameworks in this Volume are designed to install that discipline either way.

More information can be found at srjconsultingservices.com/books.

The AI Operating System™

The AI Operating System™ is a broader operational framework focused on helping organizations manage artificial intelligence with greater visibility, governance, operational discipline, security, and measurable business value. This Volume, The AI Efficiency & Process Optimization™, is Volume IV of The Operating Discipline for AI Library™, a nine-volume canon developed through SRJ Consulting & Services LLC.

Each Volume in the Library focuses on a specific area of AI governance, operational oversight, performance management, risk reduction, workflow optimization, or security strategy. Together, these Volumes are designed to help organizations move beyond fragmented AI adoption and toward structured operational management systems capable of supporting long-term business performance and accountability.

Appendix A through O: The Consulting Toolkit

Appendix A through O: collects every working instrument used across this Volume into one place, lettered to match the references in the chapter text and ordered as a consultant would work through them during an engagement. Each instrument also appears in its source chapter, where it is explained in full. The complete set is available as editable files in the companion library, ready to populate during an AI Efficiency & Process Optimization engagement.

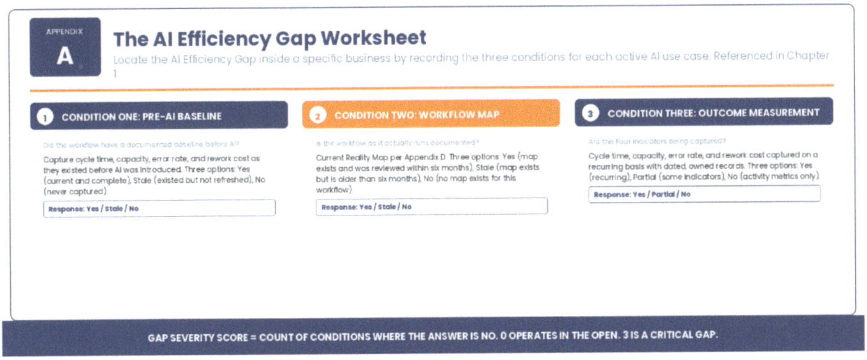

Appendix A: The AI Efficiency Gap Worksheet

The diagnostic instrument that locates the gap inside a specific business by recording the three conditions (missing baseline, unmapped workflow, activity-only measurement) for each active AI use case. Referenced in Chapter 1.

Purpose

This worksheet locates the AI Efficiency Gap inside a specific business by producing a documented record of the three conditions that combine to create the gap. Run across every active AI use case, it surfaces which use cases are operating in the open, where a baseline exists, the workflow is mapped, and outcomes are measured, and which are operating in the gap, where one or more of these conditions is missing. The output is the structured starting point that every other instrument in this Volume depends on.

When to Use

Run at the start of every 90 Day Plan cycle (Appendix K) and any time a new AI use case is introduced. The owner is the AI Efficiency lead, typically the COO or a designated process leader. Time required is thirty to sixty minutes per use case for a leadership team that already has the underlying data; longer for use cases where the data has to be reconstructed.

The Instrument

For each active AI use case, capture the use case identifier first. Workflow name. AI tool or tools involved. Owner. Date AI was introduced. These five fields give every downstream finding a fixed reference.

Condition One: The Pre-AI Baseline. Did the workflow have a documented baseline that captured cycle time, capacity, error rate, and rework cost before AI was introduced? Three options. Yes, the baseline exists and is accessible. Partial, some metrics exist but not the full four-indicator set. No, no baseline was captured. For Partial or No, note which specific indicators are missing.

Condition Two: The Workflow Map. Is the workflow as it actually runs today documented in a current Reality Map (Appendix D)? Three options. Yes, the map exists and was reviewed within the last six months. Stale, a map exists but is older than six months and has not been verified. No, no current map exists.

Condition Three: The Outcome Measurement. Are the Four AI Performance Indicators (cycle time, capacity, error rate, rework cost) being captured on a recurring basis with dated, owned records? Three options. Yes, all four captured at a defined cadence. Partial, some indicators captured but not all four, or not on a regular cadence. No, activity is being tracked but outcomes are not.

Gap Severity Score. Count the number of conditions where the answer is No (not Partial, not Stale, but No). Score: 0 (no gap), 1 (minor gap, one condition unmet), 2 (significant gap, two conditions unmet), 3 (full gap, the use case is operating entirely in the dark).

Interpretation

A use case with a Gap Severity Score of 0 is operating in the open and can be optimized using the rest of this Volume's instruments. A score of 1 indicates one specific condition that must be addressed before the use case can be measured reliably; the worksheet identifies which one. A score of 2 or 3 indicates the use case is operating in the gap and is, by definition, producing financial impact that cannot be defended under scrutiny. Score 3 use cases should not be expanded, regardless of perceived benefit, until at least one condition is closed.

Connections

The output feeds the AI Efficiency Tax Calculator (Appendix C) by identifying which use cases need their Tax estimated, the Workflow Reality Map Template (Appendix D) by identifying which use cases need a current map built, and the

Pre-AI Baseline Capture Form (Appendix F) by identifying which use cases need baseline reconstruction. A leadership team that runs this worksheet quarterly has a documented quarter-over-quarter record of how the AI Efficiency Gap is narrowing or widening inside the business.

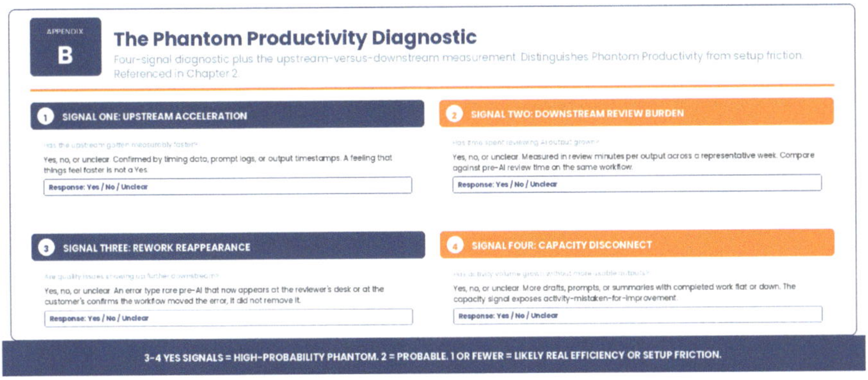

Appendix B: The Phantom Productivity Diagnostic

The four-signal diagnostic that distinguishes setup friction from Phantom Productivity, plus the upstream-versus-downstream measurement template. Referenced in Chapter 2.

Purpose

This diagnostic distinguishes Phantom Productivity from legitimate setup friction. It captures the four signals that confirm Phantom Productivity is present in a workflow and produces the upstream-versus-downstream measurement that proves it or refutes it. The output is a documented determination of whether each AI use case is generating real efficiency or generating activity that looks like efficiency.

When to Use

Run after the AI Efficiency Gap Worksheet (Appendix A) on any use case with a Gap Severity Score of 0 or 1, where the data is sufficient to test. Run any time a workflow is producing activity-level praise from team members without correspondingly improving financial outcomes. Owner: the workflow owner together with the AI Efficiency lead. Time required is two hours including the measurement window.

The Instrument

Part One: The Four Signal Check. For the workflow being evaluated, assess each of the four signals.

Signal One: Upstream activity acceleration. Has the upstream portion of the workflow gotten measurably faster since AI was introduced? Yes, no, or unclear.

Signal Two: Downstream review burden. Has the time spent reviewing, correcting, or revising AI outputs grown since AI was introduced? Yes, no, or unclear.

Signal Three: Rework reappearance. Are errors or quality issues that the upstream AI step appeared to eliminate reappearing further downstream? Yes, no, or unclear.

Signal Four: Capacity disconnect. Has activity volume grown without a corresponding increase in completed, usable outputs? Yes, no, or unclear.

A workflow that shows Yes on three or four signals has high-probability Phantom Productivity. Two yes responses indicate probable Phantom Productivity. One or fewer indicates the workflow is likely producing real efficiency or experiencing setup friction.

Part Two: The Upstream-Downstream Measurement. For the workflow being evaluated, capture across one full operating cycle (one week minimum): the upstream time (total minutes from workflow trigger through the AI step), the downstream time (total minutes from AI output through completed, usable delivery), the pre-AI upstream baseline if available, and the pre-AI downstream baseline if available.

Calculate the net change: (current upstream plus current downstream) minus (baseline upstream plus baseline downstream). A negative number is real efficiency. A near-zero or positive number is Phantom Productivity, regardless of how fast the upstream feels.

Interpretation

The diagnostic produces one of four outcomes. Real Efficiency: net time down significantly, three or four signals say No. Setup Friction: net time up but signals indicate one-time configuration issues that resolve over weeks. Mild Phantom: two signals say Yes, net time near zero. High-Probability Phantom: three or four signals say Yes, net time stable or up. Each outcome has a corresponding action documented in Chapter 2.

Connections

Phantom Productivity findings feed directly into the AI Efficiency Tax Calculator (Appendix C), where the review burden and rework cost components are derived from this measurement. They also feed the AI Process Fit Test Scoring Sheet (Appendix E) by informing the workflow fit dimension score.

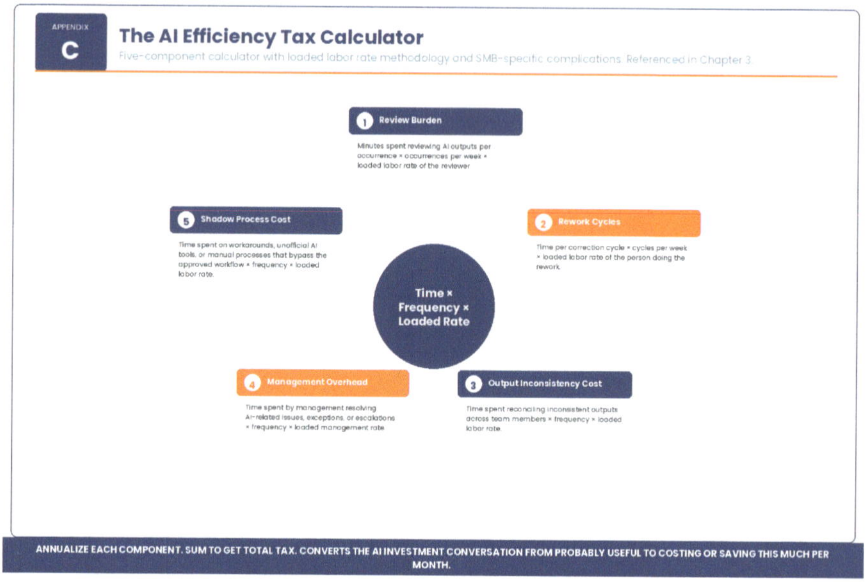

Appendix C: The AI Efficiency Tax Calculator

The five-component calculator (review burden, rework cycles, output inconsistency cost, management overhead, shadow process cost) with the loaded labor rate methodology and the SMB-specific complications. Referenced in Chapter 3.

Purpose

This calculator produces a defensible, annualized dollar estimate of the hidden cost AI is adding to a workflow through five specific components. The output is a single number per AI-supported workflow, plus a portfolio-level total across all active use cases, that can be presented to a CFO, board, or investor without falling apart under questioning.

When to Use

Run for every use case with a Gap Severity Score of 1 or lower (Appendix A), and after the Phantom Productivity Diagnostic (Appendix B) is complete. Update quarterly as part of the 90 Day Plan cycle (Appendix K). Owner: the AI Efficiency lead together with the workflow owner. Time required is three to four hours per workflow for the first calculation, one to two hours for quarterly updates.

The Instrument

For each AI-supported workflow, estimate the five components below. All estimates are annualized.

Component One: Review Burden. The time spent reviewing AI outputs before they are released or used downstream. Capture: average minutes per review, times number of reviews per week, times fifty-two weeks, times loaded labor rate of the reviewer (from Appendix M).

Component Two: Rework Cycles. The time spent fixing AI outputs after they have been released or used and an error was caught. Capture: average minutes per rework cycle, times number of rework cycles per week, times fifty-two weeks, times loaded labor rate of the worker performing the rework.

Component Three: Output Inconsistency Cost. The time spent reconciling AI outputs that differ in format, structure, or quality across team members or instances. Capture: average minutes spent on reconciliation per cycle, times cycles per week, times fifty-two weeks, times loaded labor rate. If the inconsistency is producing downstream customer-facing issues, add an estimate of customer-facing cost (refunds, credits, lost retention).

Component Four: Management Overhead. The time spent by managers and leadership team members coordinating, troubleshooting, and explaining AI usage to the team and to stakeholders. Capture: average hours per week spent on AI-related management work, times fifty-two weeks, times loaded labor rate of the manager.

Component Five: Shadow Process Cost. The time spent on undocumented workarounds, parallel processes, and informal corrections that the documented workflow does not capture but the team has built around AI failure modes. Capture: average minutes per shadow process instance, times instances per week, times fifty-two weeks, times loaded labor rate.

Total Annual Tax. Sum the five components. Round to the nearest hundred dollars for presentation. Document each estimate with the data source, the date estimated, and the assumptions used.

Interpretation

A Tax estimate below 5% of workflow direct cost is typically tolerable and reflects normal operational friction. Between 5% and 15% is significant and warrants targeted optimization (Appendix N). Above 15% indicates the workflow may be net negative on AI; the AI Process Fit Test (Appendix E) should be re-run with this finding in hand. Above 25%, the workflow is almost certainly producing negative ROI and should move to Refine or Pause status.

Connections

The Tax estimate becomes a direct input into the AI ROI Formula Workbook (Appendix H) as a benefit-side component (Tax reduction across periods). It also feeds the AI Efficiency Scorecard (Appendix G) as the financial dimension of each use case's score, and the Executive AI Efficiency Brief (Appendix I) as the headline financial figure for stakeholder communication. The loaded labor rates used in every component come from Appendix M.

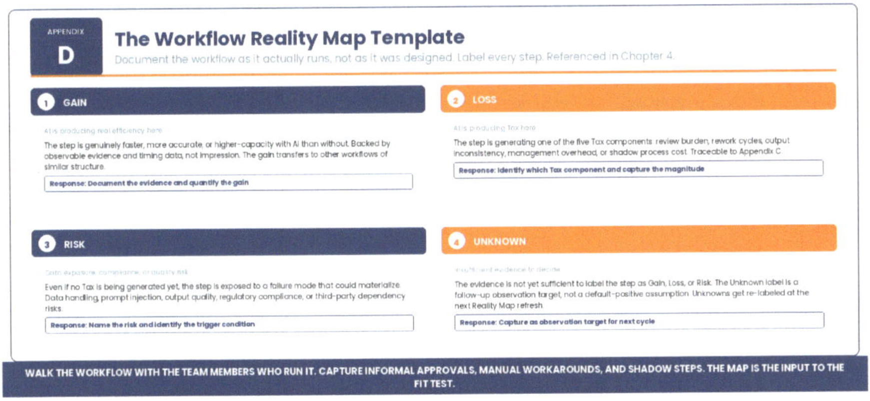

Appendix D: The Workflow Reality Map Template

The four-step mapping template for documenting the workflow as it actually runs, including informal approvals, manual workarounds, and exception handling. Referenced in Chapter 4.

Purpose

This template produces a documented Workflow Reality Map for a specific AI-supported workflow, capturing the process as it actually runs (not as it was designed to run), including every informal approval, manual workaround, exception handling pattern, and AI touchpoint. The output is the foundational reference document that every downstream optimization instrument depends on.

When to Use

Build for each of the top three AI-supported workflows in the business at the start of the first 90 Day Plan cycle (Appendix K). Refresh annually, or any time a significant change is made to the workflow or the AI tooling. Owner: the workflow owner together with the team members who actually execute the work. Time required is four to eight hours for the first map; one to two hours for an annual refresh.

The Instrument

Step One: Identify the Workflow Boundaries. Document the trigger that initiates the workflow, the final output that completes it, and the named owner. Capture: workflow name, trigger event, ending event, completion criteria, frequency (per day, per week, per month), and average volume per cycle.

Step Two: Map the Steps as They Actually Run. Walk through the workflow with the team members who execute it. For each step capture: step number and short description (one sentence), who performs it (role, not name), estimated time per step, tools used (including AI tools, by name), whether the step is documented (in the official workflow documentation) or undocumented (a workaround the team has built), and whether the step is required or optional.

Step Three: Mark the AI Touchpoints. For every step involving AI, capture: AI tool name and the specific function used, type of touchpoint (producer, where AI generates the initial output; reviewer, where AI evaluates a human output; augmenter, where AI extends a human output; replacer, where AI substitutes for a step that was previously human), and quality of the touchpoint output (high, consistently usable without modification; mixed, often requires light revision; low, requires significant rework before use).

Step Four: Label Every Finding. Walking through the map, label each step or transition with one of four findings. Gain: the step is measurably faster, more accurate, or less costly than the pre-AI version. Loss: the step is measurably slower, more error-prone, or more costly than the pre-AI version. Risk: the step contains a failure mode not currently producing loss but likely to under specific conditions. Unknown: the step's contribution cannot be assessed without further measurement.

The Map Output. A single document containing the workflow name and owner, the step-by-step sequence with AI touchpoints marked, the four-finding labels, and a summary of how many steps fall into each finding category.

Interpretation

A workflow with more Gain findings than Loss findings is a candidate for Expand status (Appendix E). A workflow with significant Loss or Risk findings is a candidate for Refine or Pause. A workflow with many Unknown findings is a candidate for measurement first, decision later. The map is a diagnostic instrument, not a planning instrument; it produces the evidence for downstream decisions rather than the decisions themselves.

Connections

The Reality Map feeds the AI Process Fit Test Scoring Sheet (Appendix E) by providing the workflow-level evidence used in scoring, the AI Efficiency Tax Calculator (Appendix C) by identifying review burden and shadow process steps, and the Optimization Action List (Appendix N) by surfacing specific steps to target for redesign.

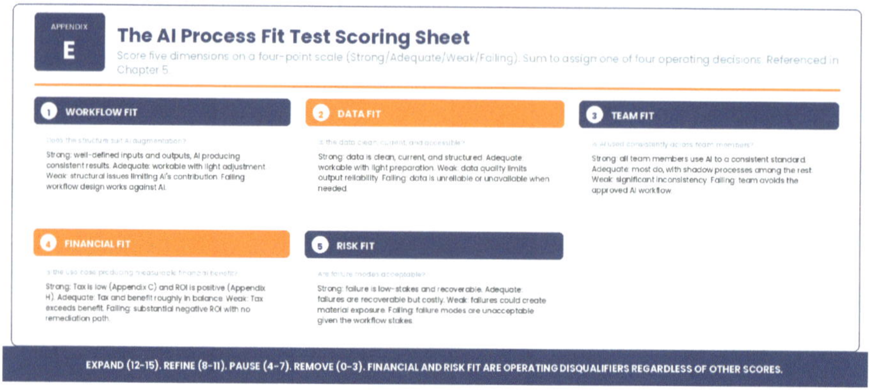

Appendix E: The AI Process Fit Test Scoring Sheet

The five-dimension scoring sheet that produces one of four operating decisions (expand, refine, pause, remove), with the practical scoring session protocol. Referenced in Chapter 5.

Purpose

This scoring sheet evaluates whether an AI use case belongs where it has been placed. It produces a five-dimension score per workflow and a single operating decision (Expand, Refine, Pause, or Remove). The output is a documented determination, supported by evidence, that the leadership team can defend if asked why a use case was kept, modified, or killed.

When to Use

Run for every active AI use case at the start of the 90 Day Plan cycle (Appendix K), and any time a Workflow Reality Map (Appendix D) is refreshed. Owner: the workflow owner together with the AI Efficiency lead, with input from team members who execute the workflow. Time required is ninety minutes per workflow.

The Instrument

For each AI use case, score five dimensions on a four-point scale: Strong (3), Adequate (2), Weak (1), Failing (0).

Dimension One: Workflow Fit. Does the workflow's structure suit AI augmentation? Strong if the workflow has well-defined inputs and outputs and AI is producing consistent results. Adequate if the workflow tolerates AI usage but produces mixed results. Weak if the workflow has too much variability for AI to be reliable. Failing if the workflow is fundamentally unsuited to AI.

Dimension Two: Data Fit. Is the data the AI relies on high-quality and accessible? Strong if data is clean, current, and structured. Adequate if data is workable with light preparation. Weak if data quality is producing inconsistent AI outputs. Failing if data is fundamentally inadequate for the AI use.

Dimension Three: Team Fit. Is the team using AI consistently and skillfully? Strong if all team members use AI to a consistent standard. Adequate if most do, with shadow processes among the rest. Weak if usage is highly inconsistent across team members. Failing if shadow processes outweigh standard usage.

Dimension Four: Financial Fit. Is the AI use case producing measurable financial benefit? Strong if Tax is low (Appendix C) and ROI is positive (Appendix H). Adequate if Tax and benefit are roughly in balance. Weak if Tax exceeds documented benefit. Failing if Tax is high and no benefit is documentable.

Dimension Five: Risk Fit. Are the failure modes acceptable given what the workflow produces? Strong if failure is low-stakes and recoverable. Adequate if failures are recoverable but costly. Weak if failures damage customer relationships or financial outcomes. Failing if failures expose the business to compliance, regulatory, or liability risk.

The Operating Decision. Sum the five dimension scores (range: 0 to 15) and apply the decision rule. Expand (12 to 15): the use case is producing on every dimension; scale to similar workflows. Refine (8 to 11): the use case is working but has weak dimensions; document specific improvements in the Optimization Action List (Appendix N). Pause (4 to 7): the use case is not producing on critical dimensions; suspend expansion, isolate the workflow, complete a structured review before resuming. Remove (0 to 3): the use case is failing on most dimensions; walk it back, document the lessons, reclaim the budget.

Scoring Session Protocol. Hold a single ninety-minute session with the workflow owner, the AI Efficiency lead, and one or two team members. Score each dimension out loud. Document the evidence cited for each score. Reach the operating decision before the session ends. Record the decision with the date, the participants, and the supporting evidence.

Interpretation

A use case can score well on dimensions one through three and still produce a Pause decision if dimensions four (Financial) or five (Risk) are Failing. Financial and Risk fit are operating disqualifiers. A use case that cannot meet them, regardless of how well the team is using it, does not belong.

Connections

The Fit Test decision feeds the AI Efficiency Scorecard (Appendix G) as the use case status, the Executive AI Efficiency Brief (Appendix I) as the headline portfolio decision, and the 90 Day Plan (Appendix K) as the prioritization input for the next operating cycle.

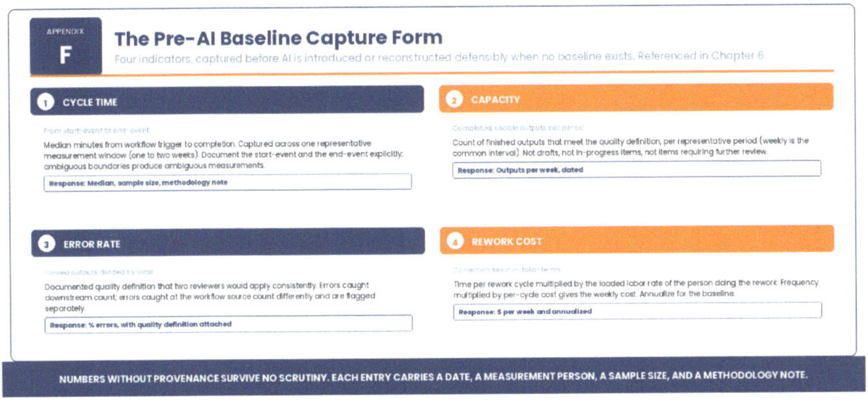

Appendix F: The Pre-AI Baseline Capture Form

The four-indicator baseline capture form (cycle time, capacity, error rate, rework) with the reconstructed baseline methodology for workflows that did not capture a baseline before AI was introduced. Referenced in Chapter 6.

Purpose

This form captures or reconstructs the baseline state of a workflow before AI was introduced, producing the four-indicator reference point (cycle time, capacity, error rate, rework cost) that every downstream measurement depends on. For workflows that did not capture a baseline at the time AI was introduced, the form includes the reconstruction methodology that produces a defensible estimate.

When to Use

Run for every AI use case identified by the AI Efficiency Gap Worksheet (Appendix A) as having a missing or partial baseline. Run before measurement work begins in the 90 Day Plan (Appendix K). Owner: the workflow owner together with the AI Efficiency lead. Time required is two to four hours per workflow for direct capture, six to ten hours per workflow for reconstruction.

The Instrument

Section One: Workflow Identification. Workflow name. AI tool or tools introduced. Date AI was introduced. Owner. Volume per cycle (per day, week, or month).

Section Two: Direct Baseline Capture. If the workflow has not yet introduced AI, or if AI was introduced recently enough that the pre-AI version is still being run in parallel for some cycles, capture directly. Cycle Time: average minutes from workflow trigger to completed output, measured across at least ten consecutive cycles. Capacity: number of completed, usable outputs per period (day, week, or month), measured across a representative period. Error Rate: percentage of outputs that required correction or rework after release, measured across at least ten consecutive cycles. Rework Cost: average minutes spent on rework per error, times number of errors per period, times loaded labor rate (Appendix M).

Each metric is dated and named. The capture document records the dates of measurement, the cycles included, and the person who captured each metric.

Section Three: Reconstructed Baseline Methodology. For workflows where AI was introduced too long ago for direct capture, reconstruct each of the four indicators using one or more of these sources. Cycle Time: time-tracking records from before AI introduction; if unavailable, structured interviews with team members who performed the work pre-AI, with consistency checks between sources. Capacity: output volume reports from before AI introduction (sales records, delivery records, document production records); compare like periods (same season, same business conditions). Error Rate: quality control records, customer complaint logs, internal correction records, audit findings. Rework Cost: combination of error rate estimate and time-per-rework estimate, sourced from team interviews or operational records.

Each reconstructed metric is labeled Reconstructed with the sources used and the confidence level (high, medium, low). Document where the reconstruction methodology has gaps the leadership team should know about when reading downstream results.

Section Four: Baseline Documentation. A single document containing the four indicators (captured or reconstructed), the dates and sources for each, the confidence level, and the date the baseline was finalized. Stored with the workflow's master file and referenced by every downstream instrument.

Interpretation

A high-confidence baseline supports unambiguous measurement of post-AI change. A medium-confidence baseline supports directional measurement but warrants caveats in financial presentations. A low-confidence baseline still produces a starting point for measurement but should be supplemented with current-state forward measurement that produces increasing reliability over subsequent quarters.

Connections

The baseline feeds every measurement instrument in this Volume: the Four AI Performance Indicator measurements in Chapter 6, the AI Efficiency Scorecard (Appendix G), the AI ROI Formula Workbook (Appendix H), and the Executive AI Efficiency Brief (Appendix I). Without a baseline, none of these instruments produces a defensible result.

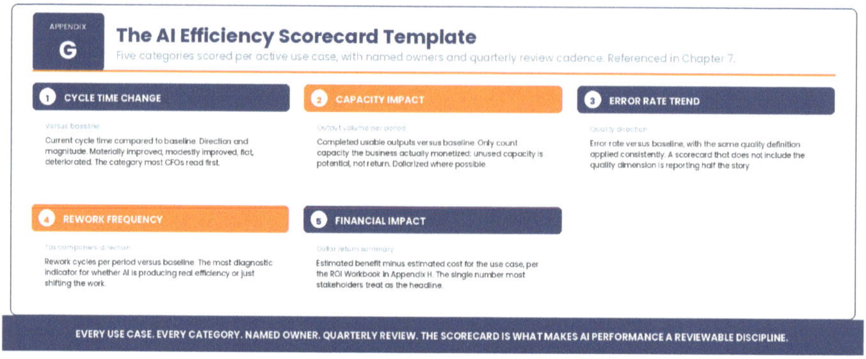

Appendix G: The AI Efficiency Scorecard Template

The five-category scorecard template with the named-owner and decision-attached review protocol. Includes the Scorecard Pattern Library for interpreting recurring score combinations. Referenced in Chapter 7.

Purpose

This template produces a single-page Scorecard per AI use case (and a portfolio-level Scorecard across all use cases) that records the operating state of each use case across five categories. The Scorecard is the central operating instrument the leadership team uses to monitor AI performance on a recurring cadence.

When to Use

Build at the start of the first 90 Day Plan cycle (Appendix K). Review and update quarterly. Owner: the workflow owner is accountable for each use case Scorecard; the AI Efficiency lead is accountable for the portfolio Scorecard. Time required is thirty to forty-five minutes per use case for the first build; ten to twenty minutes for quarterly updates.

The Instrument

For each AI use case, score five categories on a four-point scale: Strong (3), Adequate (2), Weak (1), Failing (0). Each category has named evidence sources.

Category One: Workflow Performance. Cycle time, capacity, and rework metrics compared to baseline (Appendix F) and the prior period. Strong if all four indicators are improved over baseline; Adequate if three are improved; Weak if two; Failing if one or zero.

Category Two: Output Quality. Error rate, output inconsistency, and downstream issue rate. Strong if error rate is at or below baseline and trending stable or down; Adequate if error rate is moderately above baseline but stable; Weak if error rate is rising; Failing if error rate is high and rising.

Category Three: Financial Performance. AI Efficiency Tax (Appendix C) and AI ROI Formula result (Appendix H) for the use case. Strong if Tax is below 5% of workflow direct cost and ROI is positive; Adequate if Tax is 5-15% and ROI is positive; Weak if Tax is 15-25% or ROI is marginal; Failing if Tax is above 25% or ROI is negative.

Category Four: Operating Discipline. Standardization (Appendix J) and shadow process control. Strong if standards are documented, current, and followed by all team members; Adequate if standards exist with limited shadow process; Weak if shadow process is widespread; Failing if standards are absent or ignored.

Category Five: Strategic Fit. Alignment with the business's competitive position and operating priorities. Strong if the use case supports a named strategic objective and is producing measurable contribution; Adequate if it contributes to general operating efficiency; Weak if no longer aligned with current priorities; Failing if producing distraction from higher-priority work.

The Portfolio Scorecard. A summary view across all active use cases, showing each use case's category scores, the trailing four quarters of change, the operating decision (Expand, Refine, Pause, Remove), and the named owner. Updated quarterly.

The Scorecard Pattern Library. Recurring score combinations to interpret. Three-Three-Three-Two-Three (workflow strong, output strong, financial strong, discipline adequate, strategic strong): high-performing use case; Expand. Three-Two-Two-Two-Two (workflow strong, the rest adequate): activity-strong, evidence-thin; Refine the measurement and discipline before expanding. Two-Two-One-Two-Three (financial weak): strategically attractive, financially marginal; re-run the Fit Test and the Tax calculator; likely Refine.

Three-Three-One-One-Two (financial and discipline weak): workflow and output are producing, but the operating model is not capturing the value; Refine the discipline layer urgently. One-One-One-One-One (failing across the board): Pause immediately; run the full Fit Test before resuming.

Review Protocol. Hold a sixty-to-ninety-minute Scorecard review every quarter. Walk through each use case's scores, the change from the prior quarter, and the operating decision. Document the decision and the supporting evidence. Update the Scorecard.

Interpretation

A Scorecard is operating evidence, not a report card. The point of the review is to make a decision about each use case, not to celebrate or punish a team. Use cases with weak scores are candidates for the Optimization Action List (Appendix N), not for elimination of the team running them.

Connections

The Scorecard feeds the AI ROI Formula Workbook (Appendix H) by surfacing which use cases need updated ROI calculations, the Executive AI Efficiency Brief (Appendix I) by providing the operating summary, and the 90 Day Plan (Appendix K) by identifying which use cases need optimization focus in the next cycle.

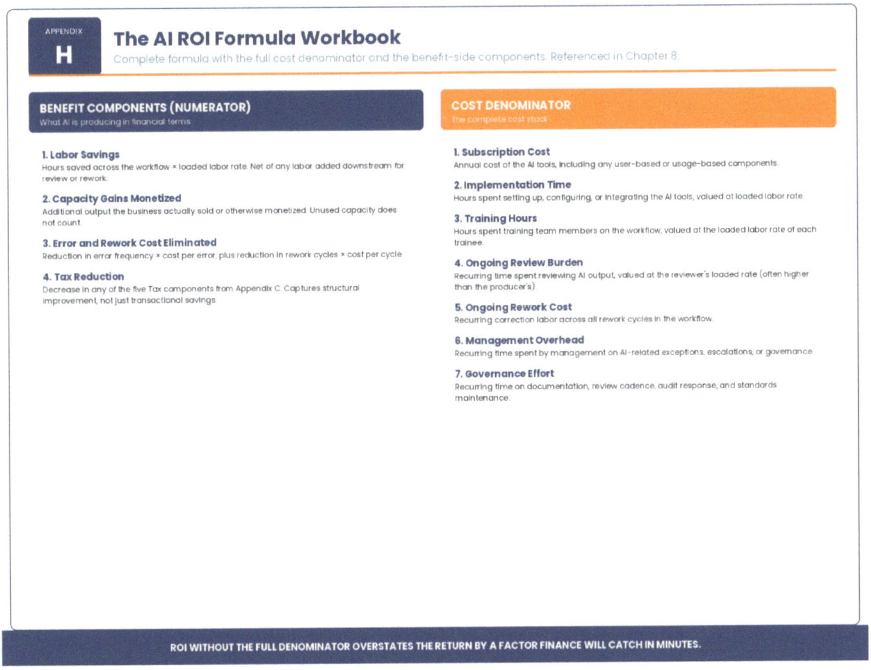

Appendix H: The AI ROI Formula Workbook

The complete formula workbook with the full cost denominator structure, the benefit-side components, and the presentation template for financial stakeholders. Referenced in Chapter 8.

Purpose

This workbook produces a defensible AI return on investment number for each AI use case and a portfolio-level ROI across all active use cases. The output is the single financial figure a CFO, board, lender, or acquirer can hold up against any AI investment claim the business has ever made.

When to Use

Calculate quarterly as part of the 90 Day Plan cycle (Appendix K). Update mid-quarter if a material change occurs (a use case is paused, a new tool is introduced, a significant Tax reduction is achieved). Owner: the AI Efficiency lead together with the CFO or financial controller. Time required is two to three hours per use case for the first calculation; one hour for quarterly updates.

The Instrument

The Formula. AI ROI equals (Total Annual Benefit minus Total Annual Cost) divided by Total Annual Cost, expressed as a percentage. Calculated per use case and at the portfolio level.

The Cost Denominator. Capture all costs annualized. Direct AI Tool Cost: subscription fees, license fees, usage fees. Implementation Cost: one-time setup cost spread across the expected useful life (typically three years for SMB AI tooling). Training Cost: time spent training team members on AI usage, times loaded labor rate (Appendix M). Integration Cost: time spent integrating AI into the workflow, times loaded labor rate. Management Overhead: time spent by managers coordinating, troubleshooting, and explaining AI usage (the Tax Component Four from Appendix C). Infrastructure Cost: any incremental cloud, security, or system cost attributable to the AI use case. The denominator is the sum. Document each component's source and methodology.

The Benefit Side. Capture all benefits annualized. Labor Savings: hours saved per cycle, times cycles per year, times loaded labor rate of the worker performing the workflow. Capacity Gains: incremental output produced because of AI, times revenue per output (if revenue-generating) or times cost saved per output (if internal). Error and Rework Cost Eliminated: pre-AI error and rework cost minus post-AI error and rework cost; if positive, captured as benefit; if negative, the workflow is producing additional error cost, which moves to the cost side as a Tax component. Tax Reduction: if the AI Efficiency Tax (Appendix C) was higher in a prior period and is now lower, the difference is benefit; this only applies to use cases that have a prior-period Tax baseline. Strategic Benefit: estimated value of capabilities AI has enabled that the business would not otherwise have; documented separately because it is harder to defend and should not be the dominant benefit component. The benefit side is the sum. Document each component's source, methodology, and confidence level.

The Calculation. ROI equals (Benefit Total minus Cost Total) divided by Cost Total, times one hundred percent. Round to the nearest whole percent. Document the calculation date, the period covered, and the named author.

Sensitivity Analysis. Before presenting the ROI number, run three scenarios. Conservative: reduce benefit estimates by 25% and recalculate. Expected: the base case. Optimistic: increase benefit estimates by 25% and recalculate. A use case where Conservative ROI is positive is a defensible-positive use case. A use case where Expected ROI is positive but Conservative is negative is a marginal use case that warrants careful presentation. A use case where Expected ROI is negative should not be presented as positive, regardless of the Optimistic case.

Presentation Template for Financial Stakeholders. When presenting ROI to a CFO, board, lender, or acquirer: lead with the Expected ROI number, show the cost denominator with each component, show the benefit side with each component and confidence level, present the sensitivity analysis results, and document the methodology in a one-page reference appended to the presentation.

Interpretation

A defensibly-positive ROI requires the Conservative scenario to be positive. A weakly-positive ROI is the Expected case being positive while Conservative is negative. A negative ROI requires either elimination of the use case (Remove decision in Appendix E) or a documented optimization plan with a quarterly checkpoint for re-evaluation.

Connections

The ROI Formula feeds the Executive AI Efficiency Brief (Appendix I) as the headline financial number, the AI Efficiency Scorecard (Appendix G) as the Financial Performance category, and the Due Diligence AI Operations Package (Appendix L) as a primary diligence artifact.

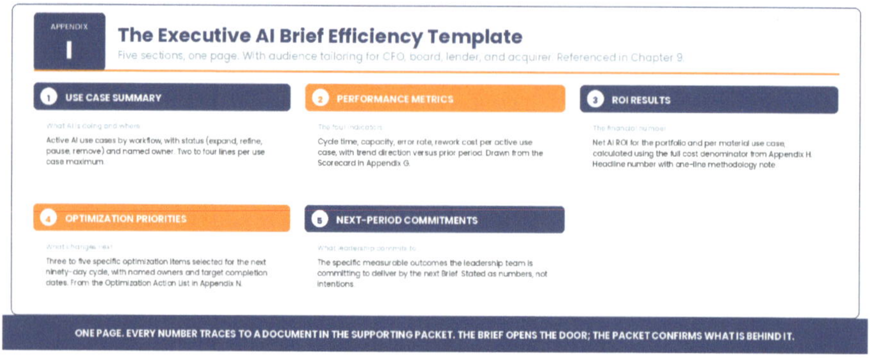

Appendix I: The Executive AI Brief Efficiency Template

The five-section Brief template (AI uses case summary, efficiency performance metrics, ROI results, optimization priorities, next-period commitments) with the audience tailoring guidance for CFO, board, lender, and acquirer audiences. Referenced in Chapter 9.

Purpose

This template produces a one-page Brief that communicates AI performance to financial stakeholders (CFO, board, lender, acquirer). The Brief replaces an AI activity update with an AI performance report. It is built once per quarter and tailored slightly per audience.

When to Use

Produce at the end of every 90 Day Plan cycle (Appendix K), for inclusion in the next leadership meeting, board pack, or stakeholder presentation. Owner: the AI Efficiency lead, with sign-off from the CFO. Time required is three to four hours for the first Brief; one to two hours per quarter for updates.

The Instrument

The Brief is one page. Five sections, each in roughly three to five sentences. No appendices on the Brief itself; supporting evidence sits behind it.

Section One: AI Use Case Summary. A one-paragraph description of which AI use cases are currently active in the business, organized by operating decision

(Expand, Refine, Pause). Captures number of use cases active, total annualized AI spend, primary workflows touched. Reads as factual inventory, not promotional language.

Section Two: Efficiency Performance Metrics. The Four AI Performance Indicators (cycle time, capacity, error rate, rework cost) summarized at the portfolio level, with directional change since the prior quarter. Each indicator gets one short sentence noting whether it moved positively, negatively, or held stable. Numbers are dated and sourced.

Section Three: ROI Results. The AI ROI Formula result (Appendix H) for the portfolio, with the cost denominator and benefit side stated. The Brief addresses the most likely follow-up question: what is the dollar figure, what are the assumptions, what is the confidence level.

Section Four: Optimization Priorities. The two or three highest-priority items from the Optimization Action List (Appendix N) for the next 90 Day Plan cycle. Each priority gets one short sentence stating what it is, what improvement it targets, and what the expected impact is.

Section Five: Next-Period Commitments. What the leadership team commits to deliver before the next Brief. Each commitment is specific, owned, and dated. Typically two to four commitments per quarter; more than that produces a Brief that overpromises and underdelivers.

Audience Tailoring. CFO Brief: emphasize the ROI calculation, the cost denominator transparency, and the Tax reduction trajectory; anticipate questions about methodology and have the workbook (Appendix H) ready. Board Brief: lead with the portfolio-level operating decisions (Expand, Refine, Pause); emphasize the change in trajectory across quarters; include the Strategic Fit category from the Scorecard (Appendix G). Lender Brief: emphasize the operating stability indicators (cycle time, error rate) and the discipline of recurring measurement; a lender wants to see that AI is not introducing operational volatility. Acquirer Brief: emphasize the documented record across multiple quarters, the operating discipline that produced the numbers, and the trajectory rather than the snapshot; an acquirer wants to see a durable operating capability, not a single-quarter performance.

Supporting Evidence. Each section of the Brief is supported by a specific artifact in the back. Section One references the AI Use Case Inventory. Section Two references the Pre-AI Baseline Capture Forms (Appendix F) and the current-period measurements. Section Three references the AI ROI Formula Workbook (Appendix H). Section Four references the Optimization Action List (Appendix N). Section Five references the 90 Day Plan (Appendix K). The Brief is one page; the evidence pack behind it is fifteen to thirty.

Interpretation

The Brief is meant to be read in three minutes by a financial stakeholder. If the Brief is not legible in three minutes, it is not a Brief. Tighten the language. Cut adjectives. Replace narrative with numbers.

Connections

The Brief is the consolidated communication output of every prior appendix. It depends on Appendix A (gap), Appendix C (Tax), Appendix E (Fit Test), Appendix F (baseline), Appendix G (Scorecard), Appendix H (ROI), Appendix J (standardization), Appendix K (90 Day Plan), and Appendix N (Action List). It feeds the Due Diligence AI Operations Package (Appendix L).

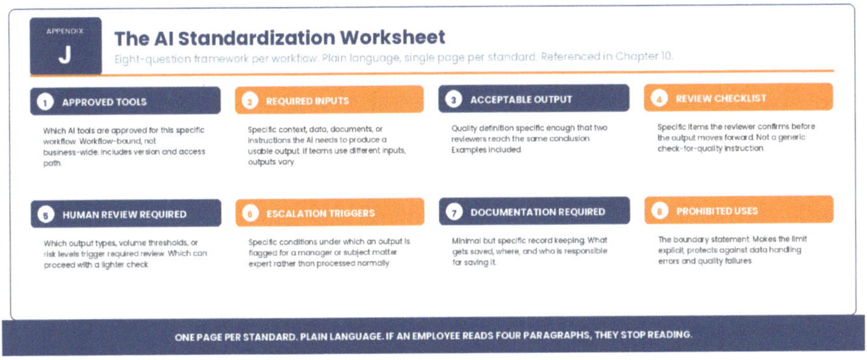

Appendix J: The AI Standardization Worksheet

The eight-question standardization framework with the shadow process review protocol and the five-component leadership alignment conversation template. Referenced in Chapter 10.

Purpose

This worksheet produces documented AI usage standards for every active workflow, captures the shadow processes the team has built around the standards, and supports the leadership alignment conversation that puts owners and review cadences in place. The output is an operating discipline that makes AI usage consistent enough to be measured.

When to Use

Run as part of the Days 31-60 phase of the first 90 Day Plan cycle (Appendix K) and refresh annually. Run any time a new AI tool is introduced or a workflow is materially changed. Owner: the AI Efficiency lead together with the workflow owners. Time required is four to six hours for the first cycle (including the shadow process review); two to three hours annually.

The Instrument

Part One: The Eight-Question Standardization Framework. For each active AI use case, answer eight questions in writing.

Question One: Which AI tool is the standard for this workflow? A single named tool. Multiple tools in parallel is shadow process.

Question Two: What is the input format the AI receives? A specific document type, prompt template, or data structure.

Question Three: What is the output format the AI produces? A specific document type, structure, or quality threshold.

Question Four: Who is authorized to use the tool for this workflow? Named roles or named individuals.

Question Five: What review step happens before the output is used? A specific check, owner, and timing.

Question Six: What is the escalation path when the AI output is unsatisfactory? Who decides what to do, with what timing.

Question Seven: What is the documentation requirement for each cycle? What gets logged, where, by whom.

Question Eight: How often is the standard reviewed and updated? A specific cadence, owner, and review date.

Each answer is documented in writing, dated, and stored where the team can reference it.

Part Two: The Shadow Process Review Protocol. Conduct a structured conversation with each team member who uses AI in the workflow. Capture: which AI tools they use that are not the standard, which inputs or outputs they produce that diverge from the standard, which review steps they skip or modify, which workarounds they have built that are not in the documentation. The shadow process review is not an audit. It is an information-gathering conversation. Team members are not penalized for shadow processes. Their shadow processes are surfaced so the standard can be strengthened or the divergence can be brought into the standard.

Part Three: The Five-Component Leadership Alignment Conversation Template. Hold a sixty-minute conversation with the leadership team. Component One: present the standards (the eight-question framework outputs). Component Two: present the shadow process findings (where divergence is occurring and why). Component Three: decide for each divergence whether to bring it into the standard (the shadow process is better than the documented standard), eliminate it (the shadow process is producing risk), or hold it (the divergence is acceptable but should be monitored). Component Four: assign owners (every standard has a named owner responsible for the standard being followed and updated). Component Five: schedule the review cadence (quarterly review of the standards, named on the calendar before the meeting ends).

Interpretation

A workflow with documented standards, named owners, and a recurring review cadence is operating in a measurable state. A workflow without these is operating in shadow, regardless of how much AI usage is happening inside it. Shadow operations cannot produce defensible AI ROI calculations because the input and output to the workflow are not consistent enough to measure.

Connections

The Standardization Worksheet feeds the AI Efficiency Scorecard (Appendix G) as the Operating Discipline category, the AI Process Fit Test (Appendix E) as the Team Fit dimension, and the Optimization Action List (Appendix N) by surfacing specific standards that need updating in the next cycle.

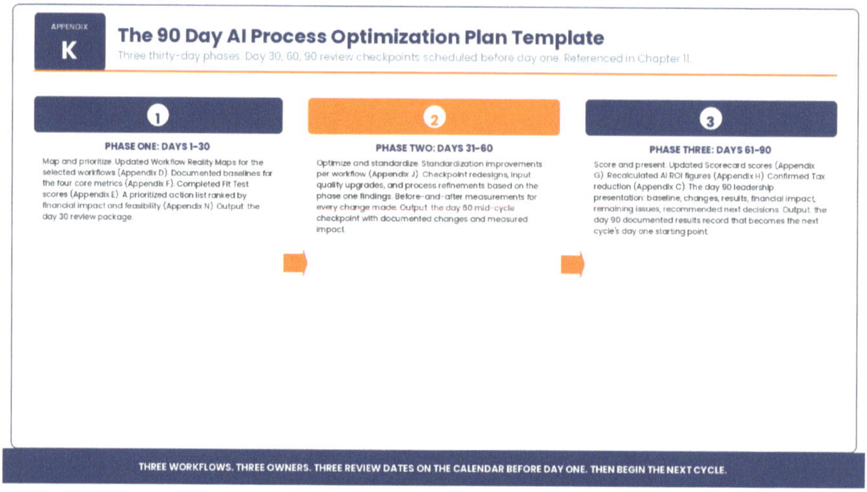

Appendix K: The 90 Day AI Process Optimization Plan Template

The complete plan template with the three thirty-day phases, the day thirty/sixty/ninety review checkpoints, and the recurring cycle setup. Referenced in Chapter 11.

Purpose

This template produces a documented 90 Day Plan for the leadership team to execute the optimization work the Volume has scoped. The Plan sequences the work into three thirty-day phases, names owners, sets review checkpoints, and produces the operating cycle that converts a one-time effort into a recurring discipline.

When to Use

Build at the start of the first cycle, after the AI Efficiency Gap Worksheet (Appendix A) has been completed for all active use cases. Refresh quarterly, with the prior cycle's Executive AI Efficiency Brief (Appendix I) feeding the next cycle's scope. Owner: the AI Efficiency lead, with sign-off from the leadership team.

Time required is half a day for the first plan; two to three hours per cycle for subsequent plans.

The Instrument

Pre-Plan Scoping. Before the Plan begins, the leadership team confirms: the three workflows targeted for this cycle (ranked by financial impact and feasibility from the Optimization Action List, Appendix N), the named workflow owner for each of the three, the named AI Efficiency lead who runs the Plan across all three, and the three review checkpoints (day thirty, day sixty, day ninety) scheduled on every participant's calendar before day one.

Phase One: Days 1 to 30. Baseline, Map, and Prioritize. Three workflows in motion. For each: Days 1-10, run the AI Efficiency Gap Worksheet (Appendix A), capture or reconstruct the Pre-AI Baseline (Appendix F), build the Workflow Reality Map (Appendix D). Days 11-20, run the Phantom Productivity Diagnostic (Appendix B), calculate the AI Efficiency Tax (Appendix C), run the AI Process Fit Test (Appendix E). Days 21-30, build the prioritized Optimization Action List (Appendix N) for the next phase, hold the Day 30 review with the leadership team, document decisions and forward items into Phase Two.

Phase Two: Days 31 to 60. Optimize, Standardize, and Measure. Three workflows in motion. For each: Days 31-40, execute the highest-impact items from the Optimization Action List, document each change. Days 41-50, run the Standardization Worksheet (Appendix J) for each workflow, hold the leadership alignment conversation, document the standards and the shadow process findings. Days 51-60, capture current-state measurements for the Four AI Performance Indicators, hold the Day 60 review, compare to baseline, document directional change.

Phase Three: Days 61 to 90. Score, Calculate, and Present. Three workflows in motion. For each: Days 61-70, build the AI Efficiency Scorecard (Appendix G) for each workflow and at the portfolio level. Days 71-80, calculate the AI ROI Formula (Appendix H) for each workflow and at the portfolio level, run the sensitivity analysis. Days 81-90, draft the Executive AI Efficiency Brief (Appendix

I), hold the Day 90 review with the leadership team, present the Brief, confirm the next 90 Day cycle's scope.

Recurring Cycle Setup. After Day 90, the next cycle's pre-plan scoping begins. The recurring rhythm: every Day 1 starts a new cycle; every Day 90 closes the prior one. The Brief from each cycle becomes the input to the next. The Optimization Action List carries forward unresolved items.

Day Thirty Review Checkpoint. Sixty minutes. Walk through each workflow's gap worksheet, baseline, and Reality Map. Confirm the priorities for Phase Two. Document any blockers.

Day Sixty Review Checkpoint. Sixty minutes. Walk through each workflow's optimization actions completed and the standardization output. Compare current measurements to baseline. Document trajectory.

Day Ninety Review Checkpoint. Ninety minutes. Walk through the Scorecard, the ROI result, and the Executive Brief. Confirm next-cycle scope. Document the operating decision for each workflow (Expand, Refine, Pause, Remove).

Interpretation

The first cycle is for installation. The second cycle is for refinement. The third cycle is where the discipline becomes durable. A leadership team that has completed three consecutive cycles has an AI operating discipline that survives leadership turnover, business volatility, and stakeholder scrutiny.

Connections

The 90 Day Plan is the operating container for every other appendix. Every other instrument is executed inside one of the three phases. The Plan produces the Executive AI Efficiency Brief (Appendix I) as its closing artifact and the Due Diligence AI Operations Package (Appendix L) as its accumulated record.

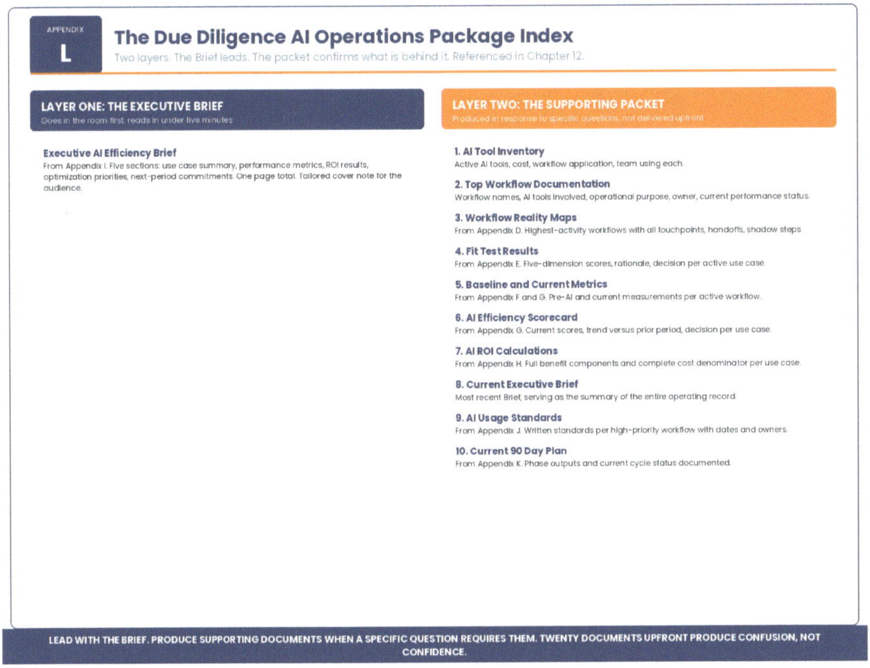

Appendix L: The Due Diligence AI Operations Package Index

The complete index of documents that constitute the AI Operations Package, organized by diligence category, with the assembly protocol for lending, investment, and acquisition conversations. Referenced in Chapter 12.

Purpose

This index lists every document that constitutes the AI Operations Package and assembles them in the order a diligence team will read them. The output is a single binder (digital or print) that converts the business's AI work into a defensible record for lending, investment, or acquisition conversations.

When to Use

Build once at the close of the first full 90 Day Plan cycle (Appendix K) and update at the close of each subsequent cycle. Refresh in full any time the business enters a high-stakes financial conversation (acquisition diligence, lender review, investment round). Owner: the AI Efficiency lead together with the CFO. Time required is four to six hours to assemble the first Package; two to three hours per cycle for updates.

The Instrument

The Package is organized into five sections. Every document is dated, owned, and current as of the Package date.

Section One: AI Operating Foundation. The AI Tool Inventory (carried forward from Volume I). The current AI Use Case List with operating decisions (Expand, Refine, Pause). The most recent AI Efficiency Gap Worksheet (Appendix A) for each use case. The named AI Efficiency lead and the named workflow owners.

Section Two: Measurement and Performance. The Pre-AI Baseline for each workflow (Appendix F), with confidence level documented. The current-period measurements for the Four AI Performance Indicators across each workflow. The most recent Workflow Reality Map (Appendix D) for each priority workflow. The most recent AI Process Fit Test (Appendix E) result for each use case.

Section Three: Financial Performance. The current AI Efficiency Tax (Appendix C) for each workflow and at the portfolio level. The current AI ROI Formula (Appendix H) result with the full cost denominator, benefit side, and sensitivity analysis. The trailing four-quarter ROI trend, if available.

Section Four: Operating Discipline. The current AI Efficiency Scorecard (Appendix G) for each workflow and at the portfolio level. The current AI Standardization output (Appendix J), including the documented standards and the shadow process findings. The current 90 Day Plan (Appendix K) and the most recent prior plan with its closing review. The trailing four-quarter Executive AI Efficiency Briefs (Appendix I).

Section Five: Governance Continuity (from Volume III). The Governance Maturity Score for each use case. The Corrective Action Register entries that affect AI workflows. The AI Risk Governance Statement. The framework alignment crosswalk (ISO 42001, NIST AI RMF, EU AI Act where applicable).

Assembly Protocol. For Lending Conversations: emphasize Section Two (Measurement) and Section Four (Operating Discipline); a lender wants to see that AI is producing operational stability, not volatility. For Investment Conversations: emphasize Section Three (Financial Performance) and Section One (Operating Foundation); an investor wants to see scalable economic logic. For Acquisition Conversations: emphasize all five sections, with particular weight on Section Four (Operating Discipline) and the trailing-quarter trajectory; an acquirer wants to see a durable operating capability. For Internal Use: all five sections in full; the Package doubles as the leadership team's operating reference.

Interpretation

The Package is the cumulative deliverable of every prior appendix and every prior 90 Day Plan cycle. A business that has assembled the Package once has the operating documentation most SMBs lack. A business that has refreshed the Package across four cycles has a multi-quarter operating record that produces a different conversation with any financial stakeholder than the alternative (verbal description and impression).

Connections

The Package consolidates every other appendix. It is the closing artifact of the Volume's work. It feeds the conversations described in Chapter 12: acquisition diligence, lender review, investment review, and internal valuation analysis.

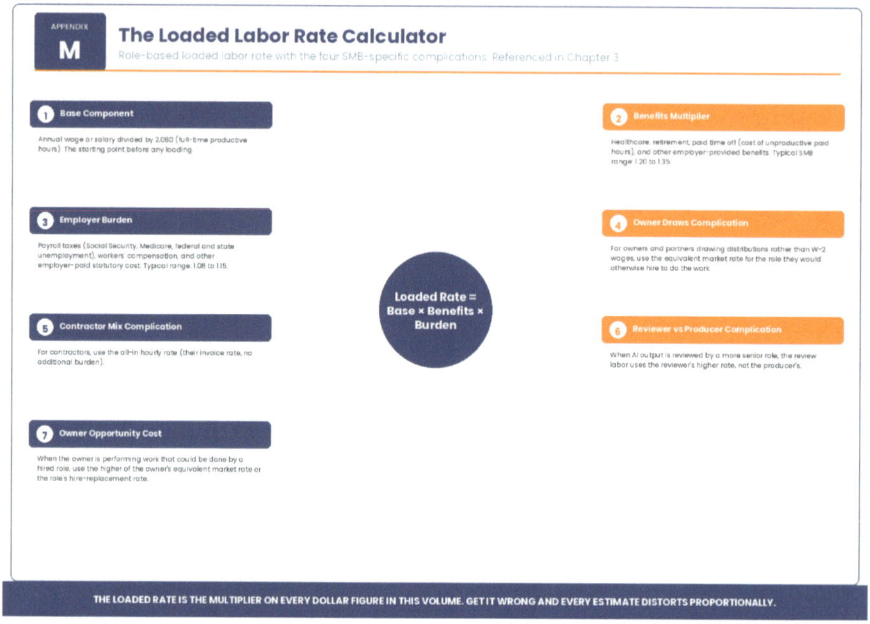

Appendix M: The Loaded Labor Rate Calculator

The role-based loaded labor rate calculator with the four SMB-specific complications (owner draws and partner equity, contractor mix, reviewer versus producer rate, owner opportunity cost). Referenced in Chapter 3.

Purpose

This calculator produces a defensible loaded labor rate per role used across every other appendix that requires a labor cost (the Tax Calculator in Appendix C, the ROI Workbook in Appendix H, the rework component in Appendix F). The output is a documented hourly rate per role that includes wage or salary, benefits, employer payroll burden, and the SMB-specific complications that standard rate calculations miss.

When to Use

Build at the start of the first 90 Day Plan cycle (Appendix K). Refresh annually, or any time a major change to compensation structure occurs. Owner: the CFO or financial controller, together with the AI Efficiency lead. Time required is three to four hours for the first calculation; one hour annually for refresh.

The Instrument

For each role used in the AI-supported workflows, calculate the loaded rate.

Base Component: Annual wage or salary divided by 2,080 (full-time productive hours).

Benefits Component: Healthcare, retirement contributions, paid time off (the cost of unproductive paid hours), and other benefits the business provides. Express as a multiplier on base; typical SMB range is 1.15 to 1.35.

Employer Burden Component: Employer payroll taxes (Social Security, Medicare, federal and state unemployment), workers' compensation insurance, and any other employer-paid statutory cost. Express as a multiplier on base; typical SMB range is 1.08 to 1.15.

Loaded Rate Per Hour equals Base, times Benefits multiplier, times Burden multiplier.

SMB-Specific Complications.

Complication One: Owner Draws and Partner Equity. For owners and partners drawing distributions or equity rather than W-2 wages, use the equivalent market rate for the role they would otherwise hire to perform the work. Document the methodology used to arrive at the equivalent.

Complication Two: Contractor Mix. For contractors providing labor inside the workflow, use the all-in hourly rate (the contractor's invoice rate, no additional burden) unless the business has reason to believe the rate is below market replacement cost. Distinguish contractor hours from W-2 hours in the workflow records.

Complication Three: Reviewer Versus Producer Rate. For workflows where the AI output is reviewed by a different role (typically more senior) than the role that initiated the work, use the reviewer's rate for the review burden component and the producer's rate for the production component. Document the split.

Complication Four: Owner Opportunity Cost. For workflows where the owner is performing the work but the work could be performed by a hired role, use the higher of the owner's equivalent market rate or the rate of the work the owner could otherwise be performing. This component is often the most significant in SMB AI ROI calculations because owner time is structurally underpriced in internal accounting.

Documentation Standard. For each role, document role title, number of full-time equivalents in the role, annual compensation (or equivalent market rate for owners), benefits multiplier and source, employer burden multiplier and source, loaded rate per hour, date calculated, and named author.

Interpretation

A loaded rate that is significantly above the team's perception of what we pay them is usually correct. The perception number is typically the base wage; the loaded rate captures the actual cost of an hour of that person's time to the business. AI ROI calculations using base wage rather than loaded rate systematically understate AI's financial impact in both directions: real savings look smaller, and real costs look smaller, producing distorted decisions.

Connections

The loaded rates feed every dollar-denominated calculation in this Volume. The AI Efficiency Tax Calculator (Appendix C) uses them for each of the five components. The Pre-AI Baseline Capture Form (Appendix F) uses them for the rework cost indicator. The AI ROI Formula Workbook (Appendix H) uses them for the labor savings benefit and the implementation, training, and integration cost components.

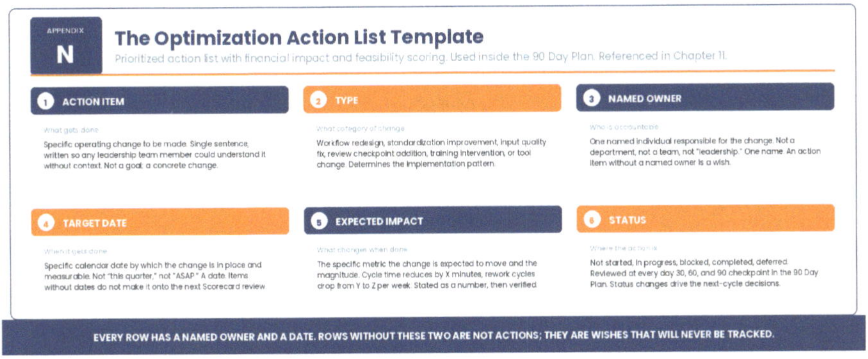

Appendix N: The Optimization Action List Template

The prioritized action list template with the financial impact and feasibility scoring methodology, organized for use within the 90 Day Plan. Referenced in Chapter 11.

Purpose

This template produces a prioritized list of optimization actions for the next 90 Day Plan cycle (Appendix K), ranked by financial impact and feasibility. The output is the executable scope of the next cycle, with each item owned, dated, and tied to a specific measurement.

When to Use

Build during Days 21-30 of every 90 Day Plan cycle as the closing output of Phase One. Update at the Day 60 review with completed items removed and new items added. Owner: the AI Efficiency lead. Time required is two to three hours to build the first list; thirty to sixty minutes per update.

The Instrument

Part One: Action Item Capture. For each candidate action surfaced by the Workflow Reality Map (Appendix D), the AI Efficiency Tax Calculator (Appendix C), the AI Process Fit Test (Appendix E), and the AI Standardization Worksheet (Appendix J), capture: action title (a one-sentence description), source instrument (which appendix surfaced this action), affected workflow or workflows (which use cases the action targets), expected impact (what measurable change the action is expected to produce, in which AI Performance Indicators), estimated cost (time and dollar cost of executing the action), named owner (who is accountable), and target completion (specific date or 90 Day Plan checkpoint).

Part Two: Financial Impact Scoring. Score each action on financial impact on a four-point scale. High (3): action is expected to produce more than $25,000 in annualized benefit or Tax reduction. Moderate (2): action is expected to produce $5,000 to $25,000 in annualized benefit. Low (1): action is expected to produce $1,000 to $5,000 in annualized benefit. Minimal (0): action is expected to produce less than $1,000 in annualized benefit (most Minimal items should be removed from the list). Use the loaded labor rates from Appendix M for time-based benefit estimates.

Part Three: Feasibility Scoring. Score each action on feasibility on a four-point scale. High (3): action can be completed within the current 90 Day Plan cycle using existing resources, with no external dependencies. Moderate (2): action can be completed within the current or next cycle, possibly with light external support. Low (1): action requires significant resources, external support, or multiple cycles to complete. Minimal (0): action is technically possible but requires major investment or organizational change.

Part Four: Priority Ranking. Combined Priority equals Financial Impact plus Feasibility (range: 0 to 6). Rank actions. 6 (High Impact, High Feasibility): do these first; schedule in the current cycle. 5 or 4: strong candidates; schedule the top two or three in the current cycle. 3 or 2: hold for next cycle; reassess after current-cycle actions complete. 1 or 0: reconsider whether the action belongs on the list at all.

Part Five: Schedule and Track. For each action selected for the current cycle, write into the 90 Day Plan with a specific phase (One, Two, or Three), an owner, and a completion date. At each review checkpoint, mark each action as completed, in progress, or blocked, with a note explaining any blockage.

Interpretation

A 90 Day Plan with three to six action items is well-scoped. Fewer than three suggests the cycle is under-utilized; the leadership team may be underestimating the optimization opportunities in the workflows. More than six suggests over-scoping; items will not complete and the discipline of the cycle will erode. The strongest cycles have three to four high-priority items completed cleanly rather than eight half-completed items.

Connections

The Action List feeds the 90 Day Plan (Appendix K) by populating its execution scope, the AI Efficiency Scorecard (Appendix G) by surfacing which categories need targeted improvement, and the Executive AI Efficiency Brief (Appendix I) by populating the Optimization Priorities section.

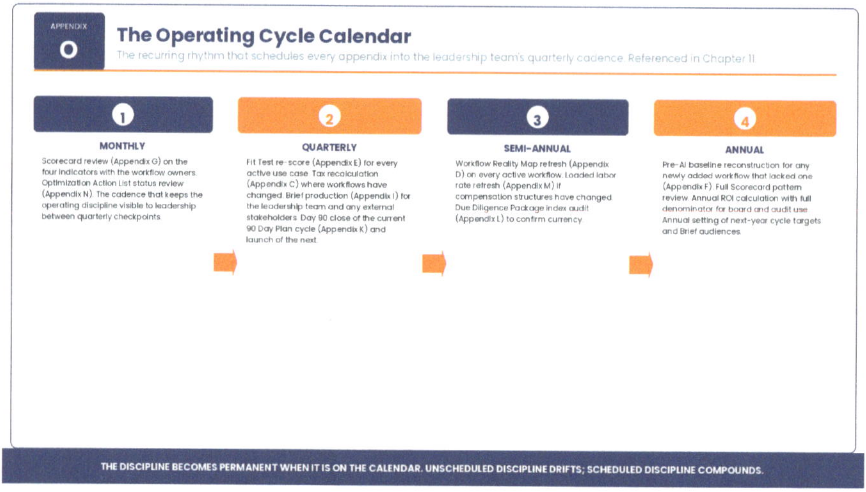

Appendix O: The Operating Cycle Calendar

The recurring operating cycle calendar that schedules the Workflow Reality Map refresh, the fit test re-score, the Scorecard review, the ROI recalculation, and the Brief production into the leadership team's quarterly rhythm. Referenced in Chapter 11.

Purpose

This calendar schedules every recurring instrument in the Volume into the leadership team's quarterly operating rhythm. The output is a one-page calendar that converts the optimization work from a discrete project into a durable operating cadence the business runs forever.

When to Use

Build at the start of the first 90 Day Plan cycle (Appendix K) and refer to throughout each cycle. Refresh annually, or any time the business's broader operating cadence changes. Owner: the AI Efficiency lead together with the leadership team's chief of staff or equivalent operations role. Time required is two hours for the first build; thirty minutes annually.

The Instrument

The calendar is organized by cadence: monthly, quarterly, annual, and trigger-driven.

Monthly Cadence. AI usage activity check (a fifteen-minute scan of which tools are being used, by whom, on which workflows); captures emerging shadow processes early; owned by the AI Efficiency lead. Action List status update (a fifteen-minute review of current 90 Day Plan items); captures blockers before they compound; owned by the workflow owners.

Quarterly Cadence (anchored to the 90 Day Plan calendar). Day 1 of Each Cycle: open the new 90 Day Plan with the pre-plan scoping conversation. Day 10: Workflow Reality Map (Appendix D) refresh completed for each priority workflow. Day 20: AI Efficiency Tax (Appendix C) updated and AI Process Fit Test (Appendix E) re-scored for each priority workflow. Day 30: Day 30 review checkpoint; Phase One closed, Phase Two scoped. Day 45: Standardization Worksheet (Appendix J) refresh completed. Day 60: Day 60 review checkpoint; measurements compared to baseline. Day 75: AI Efficiency Scorecard (Appendix G) updated; AI ROI Formula (Appendix H) calculated with sensitivity analysis. Day 90: Day 90 review checkpoint; Executive AI Efficiency Brief (Appendix I) delivered.

Annual Cadence. AI Efficiency Gap Worksheet (Appendix A) refreshed across every active use case. Pre-AI Baseline (Appendix F) reviewed and re-validated for each priority workflow. Loaded Labor Rate Calculator (Appendix M) refreshed across all roles. Due Diligence AI Operations Package (Appendix L) fully refreshed.

Trigger-Driven Cadence (run any time the trigger occurs). A new AI tool is introduced: run the AI Efficiency Gap Worksheet (Appendix A) for the workflow it touches, build the Workflow Reality Map (Appendix D), add the use case to the Scorecard (Appendix G). A workflow is materially changed: refresh the Workflow Reality Map, re-score the AI Process Fit Test, update the Tax estimate, re-baseline if needed. A high-stakes financial conversation is entering the calendar (acquisition, lending, investment): refresh the Due Diligence AI Operations Package

(Appendix L) in full. A team member departs or joins: refresh the Standardization Worksheet (Appendix J) for the affected workflows.

Integration With the Broader Operating Calendar. The AI optimization cadence does not run in isolation. The calendar should be visible to the leadership team's quarterly operating review, the CFO's financial close calendar (the Brief should be ready in time for the quarter close), the board meeting calendar (the Brief should be in the board pack on the regular cycle), and the annual operating planning cycle (the Action List feeds the next year's resource allocation).

Interpretation

A calendar that is followed produces operating discipline. A calendar that is built and then ignored produces the appearance of operating discipline without the substance of it. The point of this appendix is that the cadence is on the calendar before the work begins, and the calendar is treated as part of the business's operating commitments rather than a planning artifact.

Connections

The Operating Cycle Calendar is the temporal scaffold that organizes every other appendix into a sustained discipline. Every other instrument in this Volume has a place on the calendar. A leadership team that has built the calendar and is following it has installed the operating discipline this Volume was designed to produce.

Companion Materials

Every optimization tool, framework, and figure in this book is available online for your use, free, at srjconsultingservices.com/books/ai-business-services/the-ai-efficiency-and-process-optimization

The companion library includes:

The complete optimization toolkit as editable files. The working instruments collected in the appendices, including the Workflow Reality Map template, the AI Efficiency Scorecard template, the AI ROI Formula workbook, the Executive AI Efficiency Brief template, and the 90 Day AI Process Optimization Plan template, are available as editable files, ready to populate during an AI Efficiency & Process Optimization engagement. They are compatible with Excel, Google Sheets, and any standard spreadsheet application.

The full figure library, organized by chapter. Every diagram, framework, scorecard, and chart from this book is available as an individual file you can preview and download, ready to drop into slide decks, board presentations, internal memos, and team training materials.

All companion materials are made available without restriction for legitimate business use within the organization that purchased this book. If you are running an AI Efficiency & Process Optimization engagement in your company, share the tools with your team. If you are presenting the framework to your board, use the figures. The materials exist to make the work easier.

For additional resources, including AI operating frameworks, executive workshops, and advisory engagements, visit srjconsultingservices.com.

About the Author

Stephen R. Jordan is the founder of SRJ Consulting & Services LLC, an operator-led AI advisory practice based in Frisco, Texas. His work helps executives establish governance, oversight, and operational discipline around artificial intelligence as a managed business function, not an experimental tool.

Jordan's professional background spans accounting, enterprise technology, cybersecurity, operational systems management, and risk governance. Over the course of a career that includes leadership, consulting, and advisory roles across multiple industries, he has worked extensively in operational oversight, business systems management, enterprise security, incident response, governance frameworks, and technology implementation. His experience includes work involving operational risk management, infrastructure security, compliance oversight, and business process improvement inside complex enterprise environments.

Prior to founding SRJ Consulting & Services LLC in 2023, Jordan held senior consulting, operational, and security roles at organizations including Citi, Optiv, Intel, and McAfee. His engagements span enterprise technology, cybersecurity operations, governance processes, and systems management, with recurring focus on operational risk, infrastructure security, compliance oversight, and the management infrastructure required to keep complex enterprise environments running as designed.

Jordan earned a Bachelor of Business Administration in Accounting and a Master of Arts in Political Science with a minor in business statistics from West Texas A&M University and later completed post graduate studies from East Texas A&M University in supervision instruction & curriculum for higher education, computer science, photography, and educational media & technology. His multidisciplinary background contributed to the operational and governance-focused perspective reflected throughout The AI Operating System™ framework.

Through his work with business owners and organizational leadership teams, Jordan recognized that many companies were adopting artificial intelligence faster than they were developing the operational systems required to govern it effectively. That observation became the foundation for The AI Operating System™ and the operational frameworks introduced throughout this book and its companion Volumes.

Stephen is the author of The Operating Discipline for AI Library™, a nine-volume canon on running AI as a permanent business function. This is Volume IV, the closing Volume of Pillar I.

Today, Jordan works with organizations seeking to improve visibility, governance, operational discipline, cost control, and measurable performance across artificial intelligence initiatives and business technology environments. His work focuses on helping businesses manage AI intentionally through structured operational systems rather than fragmented adoption and uncontrolled experimentation.

For additional resources, AI governance frameworks, executive workshops, and advisory engagements, visit srjconsultingservices.com.

This page intentionally left blank.

www.ingramcontent.com/pod-product-compliance
Lightning Source LLC
LaVergne TN
LVHW071221220926
848486LV00038B/287
* 9 7 9 8 9 9 6 5 6 5 8 9 4 *